POLITICIAN, PARTY AND PEOPLE

PAGE LECTURES

Published by Yale University Press

MORALS IN MODERN BUSINESS. *Addresses by* Edward D. Page, George W. Alger, Henry Holt, A. Barton Hepburn, Edward W. Bemis and James McKeen.

(Second printing) 12mo, cloth binding, leather label, 162 pages, Syllabi. Price $1.25 net; postage 10 cents.

EVERY-DAY ETHICS. *Addresses by* Norman Hapgood, Joseph E. Sterrett, John Brooks Leavitt, Charles A. Prouty, Henry C. Emery.

12mo, cloth binding, leather label, 150 pages, index. Price $1.25 net; postage 10 cents.

TRADE MORALS: THEIR ORIGIN, GROWTH AND PROVINCE. *By* Edward D. Page.

12mo, cloth binding, leather label, 160 pages, index. Price $1.25 net; postage 10 cents.

POLITICIAN, PARTY AND PEOPLE. *By* Henry C. Emery.

12mo, cloth binding, 180 pages, index. Price $1.25 net; postage 10 cents.

QUESTIONS OF PUBLIC POLICY. *Addresses by* J. W. Jenks, A. Piatt Andrew, Emory R. Johnson and Willard V. King.

12mo, cloth binding, leather label, 140 pages, index. Price $1.25 net; postage 10 cents.

POLITICIAN, PARTY AND PEOPLE

Addresses Delivered in the Page Lecture
Series, 1912, before the Senior Class of
the Sheffield Scientific School
Yale University

By

HENRY CROSBY EMERY, LL.D.

*Professor of Political Economy in
Yale University*

New Haven: Yale University Press
London: Humphrey Milford
Oxford University Press
MCMXIII

COPYRIGHT, 1913
BY
YALE UNIVERSITY PRESS

First printed September, 1913, 1000 copies

TO
FREDERICK HALE

CONTENTS

CHAPTER I   PAGE
    The Voter and the Facts - - - 1

CHAPTER II
    The Voter and the Party - - - 33

CHAPTER III
    The Voter and His Representative - 61

CHAPTER IV
    The Representative and His Constituency   100

CHAPTER V
    The Representative and His Party - 135

CHAPTER I

THE VOTER AND THE FACTS

The foundation for these talks to students provides primarily for the consideration of the ethical side of business. It was the idea of the founder, Mr. Edward D. Page, that many members of the senior class would be shortly entering on a business career, and that the problems of business ethics should be carefully called to your attention prior to the necessity of your facing these problems in a practical way. In the series of lectures which have been given in the past few years many of these topics have been covered and the addresses published in book form. For this reason it has seemed desirable this year to depart somewhat from the original scheme and to consider certain phases of political duties and political service. It is true that the pursuit of politics as a "business" is commonly looked on askance, and yet it may fall to the lot of some of you to take up a political career as your regular occupation and, in any case, the attitude of the business man toward the ethical problems of politics is of far-reaching importance.

From one point of view a class of seniors in college needs less talking to on ethical subjects than anybody else. I doubt if anywhere a body of men could be found in whom one could put more trust to do the right thing when they see it than a group of young men who have such environment and education as yourselves. Consequently, I am not going to talk morality to you in the ordinary sense. I shall give you credit at the outset for wanting to do the right thing and for having the strength of character to do it when you know what is right. It is exactly here, however, that your need of considering this problem is greater than that of men older and less enthusiastic for the right. To most of you the problem of right and wrong seems a simple one. To men of longer experience the problem seems a very complex one. What you need most of all, then, I think, is a recognition of this complexity. You want to do right and you are starting out with a firm conviction that you will live up to a high ethical standard. Your first duty, then, is to know what is the right thing to do under given circumstances. This may sound simple to you now, but experience will teach you how far from simple such a question is. In other words, your first duty is knowledge.

I recall a classmate of mine who was hopelessly dazed by the requirement of his professor that he should write an essay on the "ultimate seat of moral authority." After I had tried to explain to him what he was expected to do, he said helplessly, "I don't see where the problem comes in. I always thought God made the laws and it was for us to obey them." This solution of the question would be as satisfactory as it is simple if the laws to which he referred were really written on tablets of bronze and specifically covered all the details of a man's public or business activity.

There is another interesting characteristic of the undergraduate in connection with the affairs of the practical world; namely, that he is inclined to look at all of them from the

4 POLITICIAN, PARTY AND PEOPLE

point of view of right and wrong. I have sometimes said that one of the greatest difficulties of a teacher is the fact that his students are too ethical. This is specially felt by a teacher of political economy who attempts to instruct a class in the actual working of economic forces from the so-called scientific point of view. By this I mean the attempt to explain facts simply as they are and to trace laws of cause and effect in a world such as this is. He is always met by the inveterate undergraduate tendency to talk in terms of what man ought to do rather than in terms of what man actually does.

We have, for instance, in economics what is known as the law of supply and demand. I think a goodly proportion of students graduate with the idea that this law has some moral force. They seem to think it a statement to the effect that prices *ought* to be determined by supply and demand, and when some case is discovered where this is not true that moral obliquity attaches to somebody. The same is true regarding most of the so-called laws of political economy, whether they be statements regarding wages, interest, taxation, or what not. I have even read answers to examination questions regarding the Malthusian theory of population to the effect that Malthus held that there *ought* to be war, famine, and vice to prevent the world from overpopulation.

This ethical coloring to the thinking of the undergraduate mind has its admirable side, but, on the other hand, it is somewhat fatal to a clear understanding even of ethical problems. One of the first essentials to clear thinking is to realize that there is a sharp distinction between the simple word "is" and the phrase "ought to be," and the important thing for our purpose is to recognize that we cannot really understand the problem of what ought to be until we recognize the fact as to what is or must be.

Apply this, now, for instance, to your first ethical duty in the field of public service, namely, your duty as a voter. After all, whatever career you follow, everyone of you will be engaged in public service. Every voter is a part of the government. As such he has a distinct moral duty to exercise his franchise in such a way as to bring about the best welfare of the nation as a whole. Again I give you full credit for being most anxious to do this. I do not need to waste words explaining to you that it is morally wrong for a man to use his vote for purposes of private gain. Being human, you will in the future doubtless often find it difficult to refrain from coloring your vague ideas of what is for the general good by your very definite knowledge of what will be for your own good. But I shall assume for the moment that your moral fibre is adequate to the strain of acting according to your conscience when the issue is clear.

Where you will have great difficulty, however, will be in determining, first, what is for the common welfare and, secondly, how this can be achieved. What I want to urge upon you is that the thoughtful consideration of these problems is a genuine moral duty for such men as yourselves, who have had the advantages of advanced education and are to be men of influence and leading in your communities.

The trouble is, the consideration of these

problems involves hard labor both in the way of thinking and of study. It is here, if anywhere, that you are most likely to be derelict in your duty to the public service. It can hardly be expected that the mass of uneducated voters should appreciate the necessity of such a knowledge, or still less that they should be able to acquire it. The indifference of a large number of the educated and prosperous class to these questions is both shocking to the moral sense and a grave danger to the state. I recognize how absorbing in these strenuous days are the demands made on the time and energy of any active man by matters of his own business and social life. Unless, however, he is willing to make a conscious sacrifice of immediate interests to an understanding of public affairs, he must expect that public affairs will be regulated by the ignorant and the unthinking.

In my own experience I have frequently found that men who belong to the so-called laboring class have given more study and thought to these questions than many who consider themselves much better fitted for their solution. And I have also usually found that it is the very men of the prosperous class who have least lived up to their obligations in this regard who denounce most vigorously the attitude of the uneducated masses and what they call the truckling of the politicians to this class of the community.

What I am trying to emphasize specially is that your ethical duty in this regard is by no means met by simply voting according to your conscience on such inadequate information as you may have at the moment. In other words, to repeat what I said at the beginning, the ethical problem is not merely the problem of voting for what is right and voting against what is wrong. Where your moral duty comes in is in coming to some intelligent conclusion as to what *is* right. You must know the facts or must have made the best effort possible to obtain them before you can meet your moral obligation as a voter.

Many acts have been passed by legislative bodies to accomplish worthy objects and have been supported with the best intentions which have, however, resulted in bringing about results exactly the opposite of those intended by their supporters. In some cases this would have been avoided by more careful reasoning; in other cases it could have been avoided by a more careful study of experience in other states or countries or in other periods of history. One of the startling facts of legislation in this country today is the frequency with which acts are passed for a certain purpose without consideration of how such acts have operated at other times or in other places. Numberless illustrations of this kind could be given if time permitted. If legislation is to be influenced largely by those who have not been trained to think clearly and have not had the opportunity to study the history of such experiments, we should expect just such a result. In proportion as it is to be influenced by men who have had these opportunities and live up to their obligations in this regard such dangers may be partially avoided.

You may feel that everything that I have said so far is very obvious, and that what I should tell you is the method by which you are to secure the necessary information to enable you to act conscientiously on measures of public policy. I wish that I could do this with any definiteness. In fact, you will find throughout these talks which I am giving a great though unavoidable lack of certainty on my part. If it were possible for anybody to mark out a definite program by which a voter or a representative or any government employee could make sure that the devotion of even two or three hours a day of conscientious labor would make all of his problems plain, there would be none of that difficulty which I have mentioned. It is because no such program can be marked out and because, whatever degree of study is given to these questions, there will always be difference of opinion, that the problem is as complex as I have said.

In brief, my purpose is not to solve your ethical problems for you, but merely to make you conscious of them. You will find me again and again stating specific problems to which I cannot myself give the answer even from the point of view of what constitutes good moral conduct. It is because I believe that much injustice is done in the world from the failure to recognize how nicely balanced many of these problems are and how necessary it is that you should never forget this in making your judgments of men and measures.

There is a vast amount of injustice done to our public men in the press, in periodical literature, and in casual conversation, through the cocksure attitude of many people that they know exactly what is right and that anybody who takes a stand contrary to their own is doing so from some unworthy motive. You will have to decide for yourselves in every case as you best may at the moment what your own action should be, but you must recognize also that other men will have to decide for themselves. If you conscientiously recognize how frequently you are puzzled regarding your own decision, you will also realize that matters which seem clear to you may puzzle others. The easy and censorious judgment of public men on the part of those who have only half thought out the problems which these men have to confront I consider one of the most immoral characteristics of the modern citizen in his relation to public service.

Unquestionably the average man—even the average educated man—derives his judgments, or rather his impressions, since usually they are not worthy the name of judgments, regarding public affairs very largely from the daily press and popular periodicals. The power of these publications in affecting public opinion can, I think, hardly be exaggerated. Some years ago the press was the all-important factor. The recent growth of the popular and low-priced magazine has introduced a new element which is largely reflected in the votes of any election at the present time. Many of you probably think in a superior way that you do not believe what you see in the newspapers. It is a somewhat popular practice among educated people to profess skepticism regarding what the newspapers say and to believe that they themselves make up their minds from some more reliable source, or through some more careful thinking, than is implied by the mere acceptance of a news-

THE VOTER AND THE FACTS 13

paper argument. If you do this I think you probably delude yourselves.

When you have seen a thing again and again in the press you forget very likely where you have seen it, but it becomes a part of your thinking and you gradually begin to make positive assertions as to what you know about something, deceiving yourselves with the idea that you have really given some independent test to the information or some independent thought to the problem. I do not know that anything else is possible in a democratic community, and I do not mean to imply that the power of the press is greater than it should be. It is easy to laugh at the old-fashioned phrase that the press is the "palladium of our liberties," but the fact remains that the free press is the greatest safeguard of democratic government.

I wish to talk perfectly frankly on this subject and can best do so probably by illustrating from my own experience. That experience put very bluntly is that newspaper correspondents or reporters are to be trusted much more than the public believes, and editors and managers much less so. Let me explain what I mean. It is, of course, true that there are some cheap reporters of the lower type who wish only to create some sensation or to make trouble for somebody; but the profession of newspaper correspondent, besides being one of the most difficult, is one of the most honorable of any profession today. In fact, from my own personal experience, I hardly know of any body of men who hold more conscientiously to the principle which the scientist is sometimes inclined to arrogate to himself, of telling the "facts as they are." The typical newspaper man has simply one ambition—to get the news. He may frequently offend the individual by the methods which he adopts or by the publication of matters which the individual would rather have kept private, but it is a part of his code of ethics that he does not perform his duty unless he furnishes the news of the day to those who wish to have it.

What makes his code of ethics high is that he wishes to publish the news exactly as it is. There are special correspondents at large political centres, such as Washington, for individual papers, and there are several great news agencies, such as the Associated Press and the United Press, whose representatives furnish news for a large number of papers, including papers of every variety of political interest. I think it may be said with certainty that both the representatives of such associations and the personal correspondents as well really wish to publish nothing but the plain, unvarnished truth. One must really have come into pretty close contact with men of this class to realize how clean-cut their code of ethics is in this regard.

You may say, then, that the conclusion is that we may probably believe whatever we read in the newspapers, and that, with intelligent men reporting the conduct of public affairs, the knowledge derived from newspapers is adequate as a basis for judgment so far as the ordinary man is concerned. The matter, however, is not so simple. In the first place, it frequently may happen that, with the very best intentions, the correspondent may get his news twisted through his ignorance of the subject which he is reporting. This is especially easy in the case of public measures involving intricate detail regarding economic regulation. I have often seen very misleading reports published regarding the opinions of some individual given in the form of an interview, or the findings of some bureau or commission, where it was obvious that the writer had acted in perfectly good faith and had intended to be absolutely accurate. The trouble was that he did not know enough about the principles involved in the matter under discussion, or enough about the past history of the measure, really to understand the true purport of what had been said. Naturally, a newspaper reporter, obliged to report the news of the day for the next morning's edition, cannot possibly take the time to make any individual study in a case of this kind. Consequently, with the most honest intentions, he may easily give out misleading information.

In the second place, you should remember that there is a difference between "news" and facts. Senator A., for instance, may make the charge on the floor of the Senate that a corruption fund has been raised to prevent the passage of a certain popular measure.

The reporter is bound to report such a piece of news. Senator A. has actually made the charge and the public is entitled to know it. It is not necessarily any business of the newspaper correspondent to investigate the charge which has been made by Senator A. This charge in itself may be true or false, but that for the moment is not a matter of news, but a matter to be settled by proper investigation. It may be that Mr. X., the great banker, gives a yachting party, on which he invites a number of well-known Congressmen. This again is a piece of news which as news is accurate, but which carries in itself no necessary implication of a desire to influence legislation to his own advantage.

It is obvious, then, that even in reading the most accurate news one should observe great care regarding the question as to how far one can draw conclusions from mere news items. Here is where the reader is likely to be mis-led through the particular policy or prejudice of the newspaper management or editor. The newspaper correspondent from Washington aims to report to his paper the news of the day as he finds it. How much

18 POLITICIAN, PARTY AND PEOPLE

of this news is to be printed, or what particular color is to be given to it, is a matter which is determined in the newspaper office itself. I do not mean to imply here any necessary dishonesty of motive, but simply the inevitable working of human nature. Newspapers and editors represent distinct and definite policies, or they represent particular parties, or in some cases particular interests. Even the most honest men may differ as to the significance or importance of different events. And we may even more expect that the men who are in the daily heat of conflict and argument must see everything and interpret everything through glasses of one color. The philosophical detachment, the freedom from bias, the fear to draw conclusions from inadequate evidence, is not for them.

Yet it is necessary that the decision as to what news should be published should be made in the newspaper office itself. What are the important facts brought out in a public hearing? What are the important statements in a public speech, and so on? I do not refer only to the formal opinions

THE VOTER AND THE FACTS

expressed on the editorial page, but to the color which is given to the facts in the news columns themselves by a choice, made in the office, of what shall be printed. It is not the business of a newspaper correspondent as a correspondent to have opinions. It *is* the business of the newspaper editor or newspaper management to have opinions and to stand for them. Consequently, however impartial the intentions of the correspondent, even his news is likely to be colored before it gets into print by the policy which the paper represents. To a certain extent, of course, the newspaper correspondent knows pretty well what kind of news is popular in the office and is, therefore, under the temptation to send only what he thinks will be acceptable there. But even if he does not yield to this temptation he cannot be certain that his whole story will be published.

It is a common saying today that nobody reads more than the headlines, and this is unfortunately only too true. The man who writes the news, however, does not write the headlines. I have sometimes maintained that the two ought to go together, but I am told that this is impossible; that headline writing is an art in itself and that, if headlines are to be made effective, they must be turned out by a master of the craft. This being the case, however, the headline becomes in itself an influence on public opinion and it not infrequently happens that the headline writer either has not fully understood the news report which he has read, or that from his own prejudice or from the known prejudice of the management, he gives it a twist which the correspondent himself never intended.

Perhaps I can best illustrate by a personal story. A representative of the Associated Press once called on me with some half information which he wished to print and asked if I had anything further to add to it. I had had much experience with him and knew him to be absolutely fair-minded and anxious to tell the facts regarding any situation as impartially as possible. I told him frankly the whole situation, which he also checked up from other sources. The next day at breakfast I read a front page column regarding the matter which irritated me extremely. The reporter came in again that day and, although I had meant to say nothing about it, I evidently showed some signs of my feeling, as he charged me with being "grouchy" and wanted to know what the trouble was. I told him it was on account of the story which I had given him the day before and which he had misused. He then took the paper off the desk and, with one hand over the head of the column, which I did not notice, asked me to read the story through again carefully and see what the trouble was. I did so and was greatly surprised. I promptly apologized and said, "What was the trouble with me? When I read that this morning I thought you had garbled it, but it reads now exactly as I gave it to you yesterday." He smiled and, taking his hand off the headlines, said, "There is your trouble. And that was as much of a surprise to me as to you." Anyone reading the story itself carefully would see that it did not bear out at all the implication of the headlines, but even I, who was more interested than anybody else, had received my impression of it from the headlines alone.

22 POLITICIAN, PARTY AND PEOPLE

Another great difficulty with trusting too much to the press for knowledge lies in what seems to be an established rule on the part of newspapers to retract as little as possible any statement which has already been made. I have heard this policy defended by newspaper men as absolutely justifiable on the ground that retractions never have any influence and that people are interested in what did happen yesterday and not in finding out that something did not happen a week ago, about which they have already forgotten. The trouble is that people have not already forgotten about the past incident. It has sunk somewhere into their consciousness and the failure to retract simply increases the influence of the first misstatement.

This is too well-known a fact to need to be discussed, but in order to illustrate what I have said as to the difference between the newspaper correspondent and the newspaper office itself, I may perhaps, at the risk of too much personal reminiscence, tell one other story of my own experience. When the Tariff Board made its first report

THE VOTER AND THE FACTS 23

a prominent Democratic senator challenged certain figures on the floor of the Senate, claiming that they had been taken from a publication of the Canadian government, but had not been given accurately as contained in the Canadian document itself. At once a leading Republican senator and a leading insurgent senator challenged this statement in turn. A hasty examination of the two documents was held by the three on the floor, whereupon both the Republican and the insurgent senator agreed that the original criticism was true. This was, of course, a vital matter if the figures offered by the Tariff Board on any question were to be considered seriously in connection with legislation. Wide publicity was given to the fact that leaders of the three different groups in the Senate had all agreed that our figures were incorrect and that a serious error had been made. This was entirely proper. It was distinctly a piece of news and correspondents quite properly reported it. The incident of the charge and the agreement of the leaders had actually occurred. Two or three leading papers, besides printing the news as it had occurred, came out with editorials saying that the incident had unquestionably largely destroyed the confidence of the public in the work of the Tariff Board.

As a matter of fact, the figures as printed in the Tariff Board report were entirely correct, and the seeming changes made from the original figures had been made in exact accordance with directions from the Director of the Canadian Census, who had first been consulted in the matter. They were necessary in order to make the Canadian figures comparable with American figures. This matter was taken up with the senators in question and with the documents before them they were all quickly convinced that an error had been made in the original criticism. Very generously they insisted, not only on amending the record, but on making a public retraction on the floor of the Senate, which was done some days later in a very handsome manner. This retraction of theirs was also news—news, I should say, as important and as interesting to the public as the original incident itself. The retraction was also

THE VOTER AND THE FACTS

reported by the correspondents to their papers as news of the day, but it was not printed. The newspaper office decided that it was not worth while, or that the incident was closed, or, in the case of those papers which had drawn editorial conclusions from the first news, that it was not wise to admit an error or make retraction. In the case of these latter papers, neither in the editorial columns nor in the news columns was any notice ever given to the fact that the original criticism had been retracted by the very men who had made it.

What I have said regarding the caution necessary in trusting to the daily press because of the extent to which information is colored by the policy of the newspaper management is equally true of the periodical press and the leading weeklies. In such publications there is practically no attempt to give the daily news as a correspondent of the Associated Press would give it. The articles are ordinarily written for a purpose and represent merely the point of view of the individual writer, or the recognized point of view of the particular periodical. There are many hack writers for periodicals of this nature who know what kind of material is acceptable to any particular publication, and are ready to present information adapted to any particular taste. Some of these writers can be well compared to what is known as a "converter" in the cotton trade. A "converter" is a man who buys goods in the "grey," which means goods woven from uncolored yarns. He then, by contract, has these goods "converted" through any of the processes of coloring, printing, or finishing of any kind to suit the taste of the customer. Much of the information which comes to the public through our modern periodicals is of this nature. The raw facts are turned out by such factories as the government bureaus, committee hearings, and so forth, and these are then artistically "converted" into such fancy products (or perhaps I should say fanciful products) as the purchaser may desire.

The conclusion from this is that the duty of the voter is to carefully scrutinize the news as he reads it, with the object of distinguishing so far as possible between the plain, unvarnished tale of the correspondent and the particular bias given to such a recital by the individual publication. I do not mean for a moment to say that the work of the editor is not of the utmost importance, or that the reader should not give due weight to the particular presentation of the facts of any journal which he trusts. What he should do, however, is carefully to recognize how far any particular journal stands for a particular policy, what its special fads are, and on what subjects it is likely to speak with some peculiar prejudice. For this purpose it is important to read several papers at the same time. I do not mean merely to read the conflicting editorials on any particular point, but to read and compare carefully the news items themselves as presented in different papers. Where the presentation of the facts is identical in several papers of different political complexion it is usually safe to accept the news as accurate. But frequently it will be found that what is supposed to be the same occurrence will be differently reported by different papers and that the statement of no one should be accepted without fair comparison with the others.

It is impossible to direct you in any formal way to sources of information on most public questions, outside of current publications. There are in many cases elaborate reports published by different bodies, articles and books by trained investigators, carefully prepared speeches of leading public men, hearings of committees, and the like. But these vary so much in value that it is almost impossible for the average citizen to know how much confidence to place in any one. One of the things most needed is a kind of clearing house of information on matters of public policy. An organization which could justly secure the confidence of the public in the way of collecting, digesting, and presenting impartially for public use the results of the best inquiries into such matters in different states and countries would be of the greatest value to the general public. Some such attempts have already been made in one or two states under an appropriation from the legislature. Probably the best illustration of this is the so-called Legislative Reference Library of the State of Wisconsin, but as yet such attempts have had only a limited and local value.

In many cases the difficulty of determining what would seem to be a simple question of fact is such as to be practically beyond the determination of the most conscientious voter. Take, for instance, the attitude of the general public toward such a measure as the Payne-Aldrich tariff act of 1909. I am not here either to defend or to criticise that measure. What I wish to suggest to you is that practically all of the statements which you have read either in defense or in criticism of it have been based on very partial information. You will hear perfectly honest men tell you, on the one hand, that this act was a revision of the tariff downward and, on the other hand, equally honest men tell you that it was a revision of the tariff upward. It would seem as if this were a simple question of fact which could be easily determined and settled to everybody's satisfaction.

I have heard professors of economics speak with complete confidence regarding this measure, as if there were no question about it whatsoever. None the less it is a question which has not been definitely settled and which can never be settled in the sense that the truth can be told about it in a single sentence. There were some striking reductions in the rates; there were some striking increases. There were many minor changes in one direction or the other. Anyone who knows anything about the subject at all knows that any attempt at expressing the change in terms of averages is meaningless. The fact is, furthermore, that, due to complicated changes in classification, it is often very difficult to determine whether a rate was increased or decreased. Frankly, the most conscientious tariff expert, if asked the blunt question whether this act increased or decreased duties, would decline to answer. If you ask him whether the duty on steel rails was increased or decreased he can tell you. If you ask him whether it was increased or decreased on cotton goods having a certain number of threads to the square inch, weighing so much per square yard, and having a certain value, he can tell you. But he would not venture any sweeping assertion. And yet the newspapers and periodicals have been filled with most categorical assertions regarding this act, and even you young men may perhaps think that you have a perfectly definite knowledge on this subject and are prepared to speak confidently in either accusatory or laudatory terms.

You may ask, in view of what I have just said, what use it can possibly be for the voter to attempt any study of these problems if it is so difficult to secure positive information. The answer is that your first duty is to recognize how difficult the problem is. I consider the attitude of the man who makes sweeping assertions regarding matters which he does not fully understand to be distinctly immoral, as it is also distinctly human.

If, then, you recognize the difficulties of many of these questions, the necessity of acting conscientiously regarding them, and yet the difficulty of equipping yourselves for an adequate judgment, you are prepared to face the next problem, which is that of the choice of the leaders whom you will follow when unable to decide each technical detail for yourselves, or the organization with which you will affiliate yourselves as best serving the public interest in the long run.

CHAPTER II

THE VOTER AND THE PARTY

In the previous lecture I pointed out that I considered the first duty of the voter in matters of public policy to be the securing of adequate knowledge. I also tried to point out some of the difficulties which confront the voter in getting accurate information. I fear that I left you in a somewhat unsatisfactory state of mind and that you felt from what I said that the problem of intelligent, conscientious action is well-nigh insoluble for even the educated voter if he is obliged to equip himself to pass upon the details of every piece of legislation. Our system of government is, however, on the one hand a system of representative government and, on the other hand, a system of party government. By representative government we mean that instead of the voters, who have the ultimate power, directly legislating according to their immediate desires, they delegate this function to representatives whom they elect by their votes and whom, presumably, they trust to represent either their own interests or the interests of the community as a whole. By party government we mean that such representatives are divided into party groups, each group representing a particular policy regarding each great public question and acting as a coherent body to carry out a party program.

Since, therefore, under our present system the voter is, in general, not called upon to vote directly upon important legislative measures, his problem becomes a secondary one; namely, that he shall choose the proper representative to voice his views or shall make the right choice between political parties to whom is entrusted the carrying out of measures of this kind. There are various problems connected with the mutual relations between the voter and his representative, the voter and his party, and the representative and his party, which I shall consider in succeeding lectures. What I want to point out here is that under a representative and party government the moral responsibility of the voter is primarily in the intelligent choice of leaders whom he will follow rather than in the acquisition of technical knowledge on each measure of public importance. Even in making such a choice, however, he can only do so conscientiously by having some intelligent opinion on at least the principles which he wishes to see enacted into legislation, even if he is content to leave the details to others. Consequently, he cannot escape the moral obligation of hard thinking and careful study to which I have referred.

This is not the place to venture on the much discussed problem of how far a representative form of government satisfactorily meets the needs of the voting population. There has been a strong agitation in recent years, as you know, in favor of substituting a more direct form of legislation by the people. Personally, I do not hold an extreme view either way regarding the problem of the initiative and the referendum. I am decidedly skeptical as to their accomplishing any very good results in the long run, but, on the other hand, I do not feel that they are fraught with very grave danger, especially in the field of local and state affairs. All I wish to point out here is the fact that under such a system the responsibility of the voter becomes much more immediate and direct and is, to my mind, largely beyond his capacity.

Direct legislation to be successful must be enacted by a body of voters who are not only convinced of some general principle which they wish to see adopted, but who have made the detailed study necessary to an understanding of the probable workings of each specific measure. The danger resulting from ignorance (whether due to indifference or to sheer limits of the human mind to handle an innumerable set of problems) is enhanced in proportion as the direct act of legislation is removed from a body of experts whose whole time and thought is, theoretically at least, devoted to these problems. In national affairs (and I am confining most of my consideration of this question to national affairs) the system of representative and party government still endures.

What, then, is the duty of the voter who is confronted with the fact that it is impossible for him to master all the intricacies of public policy in detail? In arguing recently with one of the most intelligent students of public problems whom I know, I was maintaining that the first political duty of man is to secure knowledge. To this he replied that it is the very hopelessness of even the most conscientious man getting trustworthy knowledge on most matters that made him feel that my claim was practically meaningless. In other words, he held that to advocate the impossible was to advocate nothing. His own problem for himself, he said, was to make up his mind regarding some leader whom he could trust and then follow him.

I take it that he meant some intellectual leader rather than some political leader. Such a solution may frequently work well and, of course, such intellectual leadership has been found in the past. Some of the great newspaper editors have doubtless exercised such an influence, and their readers have simply been content, once having established this trust, to follow them blindly regarding every measure. This acceptance of authority on faith is apparently less common than a generation ago. In any case, such a choice is not likely to be widely made in a community which has such an historical background of individualism and protestantism as our own.

If, instead of choosing an intellectual leader of this type, one proposes to choose a political leader, the problem usually becomes one of a choice of party. Occasionally there arises in the political world a personality who can secure for himself a following which is distinctly personal and can carry a large number of followers with him from one party to another if he chooses, and from one change of policy to another no matter how rapid these changes may be. Such extraordinary personalities, however, are not sufficiently common to change the general fact that the voter in the long run does not choose one leader, but a group of leaders; and these are not individual knights who champion nothing but their own views, but rather the leaders of an organized political body which we call a party.

There are three broad principles according to which political parties may be divided.

The first is according to sectional interest, the second according to group or class interest, and the third according to some fundamental difference of opinion regarding the principles which should be enacted into legislation for the presumed welfare of all sections and classes. The sectional interest has played a considerable rôle in our past history and, of course, at the present time the complete predominance of the Democratic party in the "solid South" rests largely on the historic grounds of sectional interest which culminated in the Civil War. It is, however, partly maintained by the question of a group interest which is still a critical one; that is, the feeling on the part of many that adherence to the Democratic party is essential to the dominance of the white population. So far no party of real power in the United States has represented group or class interest in any such conscious way as these interests are represented by the political parties in Germany. In that country, for instance, there is a party which distinctly represents the interest of the landowners, one which represents the interests of the commercial group, one which represents the wage-earning group, one which represents the interests of the Catholics, and so on. And to these are added certain smaller parties representing sectional or racial interests.

In the main, I think it may be said, although it is frequently difficult to explain what the difference is between the Republican party and the Democratic party in this country, that the division is primarily not one resulting from the clash of sectional or group interests, but is a division representing certain fundamental differences of opinion regarding the proper powers of government and the line of government policy best adapted to securing the welfare of all. That this is to be more the case in the immediate future than it was a decade ago I shall attempt to prove later.

Usually in our history we have had only two great parties and most of the voters have made their choice between these. Smaller parties may exist side by side with them, but usually secure few adherents. The average American wants to have his vote count one way or the other. To ally himself with some small, outside party is to make it impossible for his candidates to come into power or carry through their policies. In general it is the idealist and the enthusiast only who is willing to take action of this kind. This does not mean at all, however, that refusal to join such interests as these shows any lack of idealism or any lack of conscience on the part of the voter. Here is one of the questions in the decision of which the voter must search his conscience.

Take, for example, the Prohibition party. There are those who feel so strongly that prohibition of the use of alcoholic drinks is so much the most important problem of the country that they must show their allegiance to this cause by maintaining an independent political party for this purpose. Others, however, who feel as strongly on this point and are as anxious to secure the same end, decide conscientiously that the restriction of such traffic can be much better effected by voting for one of the parties which is sure to come into power and by throwing their votes to the party which will do the most toward

42 POLITICIAN, PARTY AND PEOPLE

securing their ends. The relation of the Republican party in Maine to the prohibition question is one of the most interesting illustrations of such a case.

The growing socialistic party is another case in point. Its adherents believe that the form of government which they believe in would never be adopted, or anything approaching it, by one of the established parties. The conscientious conclusion of the member of the socialistic party is that, better than to attempt some slight concession from either ruling party, is to work unceasingly for the growth of a new party which will ultimately dictate terms of its own.

These are problems of conscience which you will doubtless have to face in the future. A third or fourth or fifth party may arise at any time, which may have no possibility of immediate success, but which may represent a cause which you believe to be fundamentally just and which you think may triumph in the end through such a new organization. But two of you who believe in the same cause may make different decisions

THE VOTER AND THE PARTY

as to the best line of action. To decide to work in harmony with a powerful organization to ultimately turn it toward a certain goal is just as worthy conduct as to throw one's fortunes with what may seem a more idealistic movement by means of some new organization. It is not a question of conscience. It is a question of judgment as to the best means to the end.

If you come to a choice, as most voters do, between one or the other of the two great parties, it is here that the educated man, if he lives up to the moral duty of considering problems of public policy with care and study, ought to be able to make his choice with intelligence and a clear conscience. Having made such a choice, he must then recognize the limitations of his action. It is not for him to frame an ideal system of government or an ideal economic policy. Rather, having once cast his lot with that organization which, on the whole, he believes in, among the choice offered to him, he must recognize that now this very organization cannot accomplish even a part of the results which he desires unless it is to have the loyal support of its members. I know there are many who feel that there is something immoral in strict party loyalty. Many people of the educated class look with a superior scorn on those who work for party success year after year, even when that party follows many paths from which they have attempted to steer its feet.

I certainly believe in the independence of the voter, but I do not believe that his independence is any greater if he jumps indiscriminately from one party to the other according to some temporary feeling, or because of dissatisfaction with certain individuals. I think frequently this shows a certain lack of principle. The thoughtful and conscientious man, from his training, his historical study, his profound convictions regarding great lines of policy, ought not to be able to throw off a party as lightly as he throws off an overcoat. If he can do this, what right had he conscientiously to belong to that party before? It would indicate that he had no serious convictions, or no serious reason for his previous choice. I am speaking now of national politics and not of local or state politics. The same principles apply to a certain extent in the case of state politics and to a less extent or very slightly in the matter of municipal elections. For this reason I think it of the utmost benefit that state and local elections should not be held at the same time with the elections of a President or a Congress.

I think some of the best men in the world and the men who are striving hardest to secure improvement in public affairs waste their efforts by holding up independence as a fetish to be worshiped. The futility of many of these efforts for reform is due to the fact that reform leaders are too often unwilling to use the tools at their hands and to recognize the fact that political parties are and must be coherent, dynamic organizations. Many a man is more conscientiously loyal to some ideal who refuses to lend his support to the breakdown of an efficient organization than he who appears before the public as a champion of the very principle in which both believe.

Even at the risk of telling too many stories I venture to illustrate some of the above ideas by two instances of my own experience.

I was formerly connected with a workingmen's organization which used to meet once a week for discussion of public affairs. The club included a large variety of workingmen, from janitors to skilled mechanics, and it also included men of a variety of party and political interests. Among them one of the most intelligent, and one of the most skilled workers, was an earnest socialist.

We were invited as a club to send representatives to a special meeting to which delegates had been invited from a large number of different civic organizations for the purpose of forming some kind of federation for the advancement of the welfare of the community and general civic betterment. I read the invitation at one meeting and moved that a delegation of three should be appointed. The club was pleased to be recognized in such a movement and the vote was about to go through without contest when the socialist member arose and said that he would like to know more about it before voting in favor of the proposition. I saw at once that he was suspicious that there was some especial interest in the matter for a particular movement, and he was disinclined to give his support to anything that would not support the policies which he favored. I attempted to allay his fears by explaining that this was to be a general organization for civic betterment only, and that it had nothing to do with any particular party or any particular sect in the community; that it was simply a movement on the part of public-spirited citizens to get an organization which should stand for the best public welfare. I assured him and the others that there was no trick in it and that there was no reason why all parties should not join in it—Republicans and Democrats, Socialists and Prohibitionists, Catholics and Protestants, Jews and Gentiles.

Such an idea seemed to appeal to the others, but my socialist friend was promptly on his feet and, although remarking politely that he did not wish to oppose the wishes of "the professor" (which he always called me with a somewhat pleasant humor, in view of the fact, I think, that he was quite skeptical as to my knowledge concerning public affairs), said that he nevertheless must say a word more about the matter. Just such an organization, he said, was what he thought we ought to oppose. There was too much general agitation of this nature. "The difficulty with us is," he said, "that in these days we are always having fine meetings for speeches and for the general agitation of some general good without any definite program at all. What we need," he continued, "is to stop these 'talk fests' and get down to business. They never accomplish anything. They simply delude the public with the idea that something fine is going to be done and then nothing comes of it. The only way in which real reform is secured is for each man to stick to his organization. I believe in the man who sticks to his party and works through it. The party stands for some principle and he can only get real progress by working for the success of his party and thereby securing the success of the principle in which he believes. His party may be small and its chance of victory hopeless for the time being, but if he works

THE VOTER AND THE PARTY 49

for it and takes defeat after defeat in order to rise stronger and stronger each time he will ultimately secure the ends which he has at heart. Let us give up general meetings to which everybody can come with a clear conscience and stick to those meetings which represent a definite, clear-cut program adopted by a definite party, to be worked out by a vigorous and coherent party organization."

I was much impressed by what he said because he voiced feelings which I had felt somewhat strongly myself. Nevertheless, feeling that under the circumstances we ought to show our interest in such a movement, I got up to reply, the audience being much interested to see what answer I could make. I confess that I was somewhat put to it at first, and for a few minutes had to tread water before I could find what I thought would be a proper answer to so vigorous a statement of his position. It suddenly came over me that I had the right idea and I launched out with some confidence and, I confess, with some self-flattery as I waxed more and more eloquent (in my own mind at least), and saw that the sympathy of the audience was getting more and more with me.

My point was that I agreed with him so far as a fighting propaganda was concerned; that I agreed with him that battles were won, not by bands of music, but by bayonets and bullets. On the other hand, I pointed out, experience shows that there are emergencies in actual warfare when the band plays a very important part. It is not only an ornament and amusement in time of peace, but a real means of inspiring hope and courage for the battle and, with what I thought was a fine burst, I pointed out how, when men are disheartened and discouraged and about ready to give up in despair the band plays some inspiring air, new life is injected into the disheartened forces, the final charge is made and the day is won.

I then went on to say that that was what this new movement was for; that we were to get together and arouse enthusiasm and inspire confidence and hope and then, despite difference of opinion, we were all to go out and work with such organization as we

THE VOTER AND THE PARTY 51

believed in toward the common end of civic improvement, but that I agreed with him that it was through fighting organizations more coherent and better organized that the actual achievements were to be won. I sat down feeling that I had got out of a rather critical difficulty and much pleased with the applause that was given me.

In a moment, however, my socialist friend was again on his feet with a twinkle in his eye which foreboded that there was trouble ahead for me. "Again," he said, "I don't want to interfere with the professor and I am perfectly willing to vote for this proposition, but I must ask him just one question about that band." The moment he said that I felt my position was lost. He went on to say that he knew as well as I did that battles were sometimes won by bands as well as by bullets. In fact, he said, he knew a lot more about it than I did for he had fought in actual battles and knew what music sometimes meant. "But," he went on, "I want to ask the professor if he ever knew of a battle being won by a band that played all the national airs at once; if he ever knew of a band that amounted to anything playing 'God Save the King,' 'The Watch on the Rhine,' 'The Marseillaise,' and 'Hail Columbia' one after the other. No," he said, "bands that win battles are bands that play just one air and that air stands for a definite patriotism and a definite object—a definite something to be saved or accomplished. That is the kind of band that wins battles. And then," he continued, "when such a band stops playing does the colonel turn to the men and say, 'Now everybody run wherever he pleases'? Not a bit. He tells them to go to one point and take one battery, and that is the kind of charge that wins real victory."

You see my socialist friend was putting in very graphic form what I have said above regarding the importance of knowing to what organization one belongs, knowing why one belongs to it, and supporting it loyally. I do not wish for a moment to minimize the excellent motives of men who take the other stand, or the excellent service which they frequently do. I am only suggesting that it is quite possible that in their general kind of enthusiasm, in their genuine but somewhat vague yearning for public welfare, they will forget the importance of those organizations which are more lasting because held together by a much closer tie.

I know that many will say that the analogy is utterly beside the mark; that it is just because political parties are regardless of the common welfare and are only looking out for their own particular welfare that complete independence from them is the proper attitude for all conscientious citizens—to hold one's self aloof from all political organizations and to maintain independence above everything else. I should say to them, however, that it should be remembered that in every army there are many people of whom one cannot personally approve, there are frequently leaders who are inspired by personal advantage in the way of glory or otherwise, but that armies in which loyalty to the organization is not fundamental are seldom successful. One is reminded of Macaulay's phrase—or was it Bagehot's?—that many battles have been won by a bad general, but that none has ever been won by a good debating society.

It is also true that there have been in the past military organizations that did not stand for any high patriotic purpose, or for any common cause except personal plunder and aggrandizement. But the way to meet such organizations, whether in the past or today, would seem to be to form some other organization equally coherent and well-organized, but standing for some higher purpose. In other words, when a man becomes convinced that any one particular party or any two parties do not in their essence and at their best stand for *any* policy which makes for the public good, the question is whether the best answer is not to organize a new political party and fight the old ones with a higher purpose but with the same recognition of the power of organization.

Please understand that I am not making these assertions as positive statements of the one method to be pursued. I am presenting the problem to you for your consideration. If you were men of a somewhat different class in society or of somewhat different character, I might emphasize the importance of the other point of view. As I have already said, I believe that the men who are afraid of all party allegiance frequently do much good, but I believe it is exactly the college men who over-emphasize the value of this freedom and fail largely to appreciate the value of organization. It is because I think you are more likely to make the error in that direction that I am perhaps speaking somewhat excessively on the other side. In any case you will have to make your choice, and the choice can be made honestly and conscientiously either way.

Here again is just the difficulty of which I spoke before. It would be easy to choose if one only really knew what were the best. One's decision will probably be largely a matter of temperament, but at least you should recognize that it is a real problem, that men can differ in their opinions about it, and that because another man chooses differently from you it does not mean necessarily that he has a lower standard of morals or a lower degree of intelligence.

Just one other story to illustrate what seems to me the danger of futile action on the part of well-meaning men who consider that their conduct is more moral than if they acted otherwise. I was invited once to a small meeting of representatives from two different and worthy organizations, both working for the public good within the community and in certain matters having much the same objects in view.

They were both large and powerful organizations and each maintained a committee regarding this or that matter of public improvement, each of which acted independently from the other, made its own investigations, its own recommendations, and carried on its own campaign. It was thought by the head of one of these organizations that it was very desirable for the two to come together and stop such duplication of work. It happened, however, that the president of one of these organizations, who was one of the delegates at the meeting, was the local boss of one of the political parties. This particular organization which he represented was non-partisan and his position as political boss had nothing to do with his relation to it. The representatives of the other organization were less of the business type, but men of the highest scholarly and social standing, and no more earnest and conscientious men could be found in matters relating to the public good.

After much discussion it became evident that they were not willing to meet approaches from the other side, or to give up their independence in any way, and that the negotiations were very sure to be a failure. The politician of whom I speak saw the position very soon and said frankly that he realized that the difficulty was that they were suspicious of him because he was a politician and he knew that they felt that a boss was a dangerous man to deal with; that somehow he would turn the organization to his own purposes. Now I am not saying here who was right in this case. Possibly this was a case where independence was essential and where it was dangerous to come to any mutual agreement. I only suggest that this unwillingness on the part of the reformer to deal with a certain type of man is as likely to appear where the intentions of the politician are really completely disinterested so far as the matter in hand is concerned as would be the case where he had some particular axe of his own to grind. There is such a prejudice among certain classes regarding political organizations and politicians that many a group of worthy citizens would simply instinctively react in such a case and refuse to have any dealings under such circumstances.

But the answer the politician made was this: "I may be a boss, but that is my own business. I have never asked for any office and if I like to run politics and control a political organization I consider that a perfectly honorable object. In any case," he said, "I do not care to discuss that proposition further." He claimed that in these particular matters, which had to do with the health and welfare of the town, that, regardless of his political affiliations and merely as a citizen, he was as anxious to have the thing "done right" as anybody else. "And now," he said, "the proposition is this. You men, I recognize, know more about these things than I do. You are better educated, you have traveled more, you have better facili-

THE VOTER AND THE PARTY 59

ties. But," he said, "it won't do you the slightest good to get together and talk these matters over. If you are going to accomplish your purpose you must get action, and action must be taken by the government. Now, do you want action, or do you want independence in talk? If you want the latter, go on talking for twenty years and satisfy yourselves and I won't bother you, But," he continued, "I am the fellow who can get things done. If you will agree to figure out what ought to be done I will take care of the men who cast the vote, who never would understand you, and between us we will get both intelligence *and* action." And with that he left the conference and he also left the conferees entirely unconvinced.

Here again, you see, were two concrete moral problems; first, what are the best methods to get immediate and concrete results; second, what are the far-reaching effects which may result from a conclusion to use the best means at hand. These problems will meet you constantly in the future, and again I remind you that it is not simply an easy question as to what is right and what is wrong. They are questions for the most careful consideration, on which men of equal integrity can differ strongly. I only suggest again that the danger to men of your type is likely to be that you will act really immorally when you think you are acting morally; that sometimes when you feel a spiritual fervor because of your independence and the complete cleanness of your skirts you may really have committed the immoral action of blocking some genuine good through some petty pride in not being willing to use any tools except those of your own framing or of your own choice.

CHAPTER III

THE VOTER AND HIS REPRESENTATIVE

The question which I wish to take up with you today is the duty of the voter regarding the choice of his representatives and especially his particular representative in Congress. I frequently hear it said by men who take great interest in public affairs and study the records of candidates with care that they always vote for the "best man," regardless of party. This sounds very sensible from one point of view and also seems to show a fine independence of dictation from any organization. On the other hand, it has been maintained by many leading moral thinkers for many years that the fundamental principle in the case of popular government is the principle of "measures, not men." You see, it is very easy to get two well-sounding phrases, each of which seems to illustrate a moral principle to be followed and each of which is opposed to the other.

I find personally among my acquaintances that there are many who are interested in legislative measures and pay a good deal of attention to their consideration and yet who, when they come to make up their ballots, seem to think only of the relative merits of the individual men who are candidates, and to give little thought to the more far-reaching problem of the ultimate effect of their vote.

In an address to his constituents at Edinburgh the astute Macaulay discusses the question of measures versus men in a very interesting way. He quotes with approval the principle of politics that support should be given to measures, not men, but then goes on to consider cases in which one must also consider the problem of personality as well. His particular point, as I recall it, has to do with the question of the way in which laws will be administered by one group of men as compared with another group. This is, of course, important. The enforcement of legislation is as important as its enactment. To

THE REPRESENTATIVE 63

vote at one and the same time in favor of a certain legislative policy and for a group of men who will not enforce the very policy advocated is practically to nullify the first vote altogether. He points out that under a certain group of administrators a bad law may be handled in such a way as to do no harm, whereas a law which seems admirable on its face and has been most carefully framed may be treated in such a way as to fail entirely in its results. Such occasions may perhaps arise, but I am not concerned with that particular problem at the moment.

In general we may take it for granted that measures and men go together; that men who stand for certain measures will enforce them as well as enact them. Under such circumstances, if one is dealing with broad public policies affecting the general welfare, the only way in which to secure the adoption of one's own particular policy, or the nearest approach to it possible, is to vote for the men who stand for that policy. You may take this to mean that I advocate complete party regularity and hold that the voter has no need even to scrutinize his ballot, but simply to mark the party of his choice. Let me try to explain to what extent I think independence regarding individual candidates can go together with the conscientious belief in the support of measures rather than men.

In the case of local administration, for instance, party measures are by no means so paramount as in national affairs. There is likely to be much less holding to strict party lines in a board of aldermen than in Congress. In municipal elections also the question is frequently not so much a question of some policy of party government as the mere honest and efficient administration of municipal affairs. For a large part of these affairs the question of party allegiance is as little important as it would be in the choice of the officials of a corporation. The result is that the principle of men rather than measures, I think, applies primarily here.

This is one of the reasons why it is so important to separate local elections from national elections in order that the voter may make his choice independently in the two cases. Possibly some cases may arise in which important problems of local policy of a general nature are involved on which the parties take opposite sides with definite and clear-cut programs. In such cases, where the voter considers this question of policy to be the important feature of the election, his choice must be for the man who supports the measure without much regard for his other qualifications, but, as I have already said, I am more concerned in these talks with national affairs than with local or state affairs.

In the case of state elections the situation is more or less half way between that of municipal elections and that of national elections, and the voter must choose his ticket with unusual care. Where the important questions involved are questions of mere administration the municipal rule would hold. Where there are questions of state policy the conscientious voter would, however, in many cases elect a man whom he considered inferior in capacity, or even in conscientious devotion to public duty, if he is sure of his vote on a particular issue. It is true that in state affairs again the party regularity of the legislator is not so compelling as in national affairs, and one might in some cases feel that the best legislation would be secured by electing the best man. Nevertheless, the situation varies from state to state and in many states party organization in state affairs is still very rigid and questions of state policy, some of them of the utmost importance, are determined very largely by party votes.

The voter must also remember that frequently the seemingly high-minded legislator who talks about what he will do when he gets to the legislature actually and of necessity is there coerced in a case of emergency to stand by the party in all questions of party measures, and it by no means follows that the voter will secure the results which he desires by voting for the candidate whom he considers the ideal man. It should be remembered that a legislature must be something more than a collection of model citizens. It must be a collection of men with views, men with a program which they intend to carry through. The problem here, as in national affairs, is more a problem as to what one thinks of the leaders of a given party than of what one thinks of the rank and file.

The result is that the choice is often a very difficult one for a conscientious man who must frequently vote for a man of whom he does not personally approve simply because he does approve of the leader whom this particular candidate will obey. I very frequently hear men say that, in making up their ballots for an election where, for instance, there are two representatives from the same district, they have voted for one Republican and one Democrat, and speak of this with pride as showing careful scrutiny of all candidates independent of party and a vote absolutely according to conscience. But if measures of real importance, on which parties are divided, are to come before such a legislature, I can see no particular reason for self-congratulation on the part of a man who has cast his vote in such a way that it will become absolutely nullified on the final result,—that is, in voting for one candidate who will vote yes on a particular measure and for another who will vote no. And yet many citizens fail to recognize that they may be nullifying their action in such cases.

There is another factor in the case, so far as state and local elections are concerned, which should not be forgotten. As our politics are now organized, the party divisions are, nominally, at least, the same for local affairs, state affairs, and national affairs. This is both unfortunate and complicating. If, for instance, the Democratic organization in a given city is closely connected with the state organization and that in turn with the national, the voter must face the question of how the effect of his vote will be felt in the wider field of national politics. A local election may come on the eve of a great and perhaps very doubtful national election. The voter may feel that the local "machine" of his own party ought to be disrupted, but this disruption might mean a loss of party strength at the national election in which he whole-heartedly supports his party. If he votes on the immediate question of good local government alone, he will vote against his own party. If he is primarily concerned with the success of a certain program in national affairs he may conscientiously decide that for the moment the problem of good local government must be sacrificed to some more pressing and far-reaching issue. In fact, such problems do present themselves continually.

Unfortunately, it is this very fact which helps to maintain an undesirable local "machine" so long in power. Pressure is constantly brought to bear on the best citizens not to endanger the party strength in the wider field. Just when this pressure must be yielded to, and when opposed, is a question which needs the deepest consideration. I have already said that the problem is rendered easier when local and national elections are held at different dates. It will never be solved, however, till we have different party *names* in local and in national affairs. The citizen of London, for instance, chooses between the "Conservative" and "Liberal" parties when voting for Parliament. When voting for the county council, however, he chooses between the "Moderates" and the "Progressives." It would be a great help to honest municipal government in this country if the names Republican and Democratic should disappear from local ballots altogether. This seems to have been accomplished in such communities as have adopted the so-called commission system of government, where there is presented to the voter at election time the whole list of candidates for choice, each standing on his own record regardless of party affiliation.

Coming now to our main question in this lecture, of the relation of the voter to his national representative, I think we find that in such cases the problem is usually far more a question of the party to be supported than the particular man to be elected. It is true that there may be cases where it is absolutely essential to punish some individual for flagrant misconduct, where the moral issue as to what type of man a district will allow to represent it becomes for the moment more important than the question of what measures are likely to be enacted for the public good or the public harm, as the case may be. Such a case, however, should be perfectly clear in the mind of the voter and should not be merely a matter of prejudice, of pique, or

THE REPRESENTATIVE

even a feeling that a man is not of sufficiently high intellectual and moral calibre to be a member of great influence in national councils.

The voter must recognize the possible far-reaching results of his action; that by means of a single congressional vote the whole course of policy may possibly be changed. To risk such a change of policy lightly and without due consideration is not truly moral conduct, even if it may seem to be more moral at the moment to refuse support to a particular individual. Consequently, to accomplish one's own desire for the best public welfare a degree of party regularity becomes here more essential than under other circumstances. The question is, Who control or direct the party? And the answer is that the party is, for the moment, controlled or directed by a small group of men who dominate it in legislative affairs and that the importance of the individual representative is relatively not great.

Where you must search your minds is primarily on the matter of whether or not you have confidence in the leaders of a given party. If I may use the military figure again, it is a question of the choice of general rather than of private. If you believe that for the success of your cause a certain general needs all the support possible, you would then try to provide a private to help fill up his ranks. You would not simply pick out the man you thought was the best soldier, provided the soldier himself had the choice of which general he would follow. If there were only one man available to send to your general you would send that man. The other man might be a better soldier, but you would not accomplish your end if you were to have him sent to the other general.

Take the case of the two parties in Congress. If you had a direct vote for the leaders for whom would you vote? Would you vote for the three or four Republican leaders, or for the three or four Democratic leaders, if you had your choice? It is curious how little the importance of this problem is recognized by many very intelligent people. I have been often taken to task by friends for voting for a certain Congressman on the grounds that the rival candidate was a much abler man and a much "better fellow." I have explained that I did not cast my vote for representatives according to my personal judgment of the two men, but that I cast it according to the policies which I wanted to see enacted. Even then I have been accused of taking a low attitude in not choosing the best representative judged purely from the personal point of view.

In the same way you have probably often heard men say, "I am not going to vote for so and so in this election. There is nothing particularly against him, but he is narrow-minded, partisan, and will never make headway in Congress. Who cares in Washington what he has to say on any question, or what influence can he have?" Well, the answer is that he can have the influence of his vote and that is about all the influence that any of them have. In nine cases out of ten the representative has little to do but follow his leader. More important than the slight glamor which comes from having a representative from your district who can make a fine personal impression or an eloquent

74 POLITICIAN, PARTY AND PEOPLE

speech is the need of having a man who will support the general in whom you believe.

Before going farther, let me make two observations which I hope you will keep in mind in connection with what I say later as well as now on this matter of the relative power of the leaders, and the relative insignificance of the rank and file. First, I do not wish to exaggerate the power of those party leaders who are themselves members of the House of Representatives, although I am dealing in this lecture primarily with the voter's relation to that branch of Congress. The leaders who direct the legislative program of Congress are frequently, of course, members of the upper House. They may also be entirely outside Congress and in fact holding no office whatsoever. At the present time (May, 1912), for instance, the acknowledged leader of the Democratic majority in the House is Mr. Underwood of Alabama, but there are members of the party, who for the time being are private citizens, whose influence in determining party policy, and even the direct action of the members of Congress themselves, is very

THE REPRESENTATIVE 75

great. The influence of Mr. Bryan, for example, on the attitude of Congress itself is perhaps as great as that of any leader in either House or Senate. This does not, however, alter the main fact which I have stated; that one's choice of the individual representative must be determined, not so much according to one's judgment of the individual man, as according to the group of leaders who, for the time being, he is practically obliged to follow. Furthermore, the influence of outside leaders on members of Congress will be in many cases felt indirectly through the leaders of the floor itself. This means either that the outside influence acts directly and sympathetically on the leaders on the floor, or else that these latter are forced to recognize the power of such outside influence and to compromise with it in order to maintain their own leadership.

This suggests the second point, which is likely to occur to you. After all, do the leaders determine the policy of the party, or does the rank and file of the party determine its own policy and dictate to the leaders? I have seemed to imply that the former is the case rather than the latter. In doing so, however, I have in mind short periods only. It is a matter of mutual action and reaction between the leaders and the rank and file, and, in the long run, of course, it is with rare exceptions impossible for any leader or group of leaders to swing a whole party away from its established principles. Those men rise to leadership who represent the long-run opinion of the rank and file. In this sense the party dictates to its so-called leaders.

The psychology of political leadership is peculiarly fascinating and is frequently misunderstood by those who look on every compromise as a sign of weakness. Among the many intricate problems of legislation which Congress has to face there will inevitably arise frequently cases where the acknowledged party leader will be forced to support certain measures in which he does not wholly believe, or to withhold his support from measures with which he is in hearty sympathy, simply because he knows that he cannot carry his party with him. To

THE REPRESENTATIVE 77

act otherwise, it seems to him, is to endanger not only his personal leadership, but his ultimate influence for what he considers the national welfare, without accomplishing any good results as an offset to such a sacrifice. I have often heard men speak somewhat sneeringly of such conduct and assert that a man who recognizes this necessity cannot really set up a claim to qualities of genuine leadership. This seems to me a mistaken attitude. A man who does otherwise cannot lead in the field of politics. The best swords are not those which are most rigid. Lowell wrote of Lincoln's mind "bent like perfect steel to spring again and thrust." The old saying that you can lead a horse to water, but you cannot make him drink, must frequently be paraphrased in the mind of an astute leader to the effect that there is no use in leading a horse to water if you cannot make him drink. Of course, in matters of fundamental principle, or in the face of a grave crisis, the true leader will stake his all on the effort to swing a reluctant party toward his own point of view; but many occasions of lesser importance will arise where he must recognize the limitation of his power over his followers.

You will see, then, that I make very decided limitations to the proposition that the leaders make the party program. But again, it seems to me that these limitations do not alter the fact on which I have insisted; that in voting for the ordinary Congressman one should always recognize that one is voting for a private who, for the time being, will take his orders from the constituted leader. The fact that the great body of the voters ultimately determine party policies, or that the rank and file in Congress itself sets a limit to the power of the leaders, does not alter the fact that in nine cases out of ten the man for whom *you* vote will be forced to follow a program which has been prescribed with slight regard to his individual point of view.

Men who always vote according to their opinion of the rival candidates do so, I know, from most conscientious motives, but I also know that they think that men who do the opposite are hide-bound partisans. What I am suggesting is that such conscientiousness

THE REPRESENTATIVE 79

is frequently due to a superficial consideration of the ultimate results; that it is based on a lack of due knowledge of the situation; and that just in so far as it is due to superficiality or carelessness it is not highly moral action.

Of course, there is the danger of a still lower attitude and yet a not uncommon one, especially among men who pride themselves on their independence. This is the danger of voting for a man because you happen to like him, because he belongs to your crowd, because he is a good friend and neighbor, because he is one of your group at the club, and so forth. It seems almost unnecessary to point out the immorality of such conduct as this where issues of vital importance are at stake, and yet I have seen enough of it to know that you will frequently be tempted toward such action in the future. I have heard one reason given for nominating a certain man as candidate that he probably would pull a big vote with such and such an organization, meaning thereby a social organization in which he was popular. Unfortunately there is too much voting of this kind, which I consider distinctly immoral.

It is not always easy to take the other attitude. Any honest voter ought to be perfectly willing to say exactly for whom he is going to vote or for whom he has voted. But if you are on intimate terms with a man at your club you may feel embarrassed to decline to give him your support as against a man whom you do not know at all, or perhaps one whom you consider much inferior in intellect and character. Of course, the politician does not have this feeling at all. He can dissociate his friendships from his politics with perfect ease. It is unfortunately the good-natured and well-meaning amateur who is most likely to feel this pressure—the man who perhaps in most matters has a higher standard of morality than the average. It is easy, however, to use here the false cloak of independence and freedom from party as an excuse for easy-going friendliness and good fellowship.

One other point I should like to bring out in this regard is the relation of the representative in Congress to the President. I shall speak later of the growth of the initiative of the executive in legislative matters and its significance. I only wish to mention in this connection the attitude of the man who, in following out his theory of independent voting, casts his vote for a President of one party and a Congressman of the other party. I have found this to be a not uncommon practice, especially among the educated classes. It is a practice due to the theory of voting always for the best man. You can see at once, however, that the result of it is for a man in large measure to nullify the effect of his vote. Assuming for the moment (rather than to anticipate the later discussion in detail) that you expect a President to have a definite program intended for the benefit of the whole country, it still remains true that he can best carry out his program by means of his own party. In such a case to vote for a Republican President and a Democratic Congressman is to say virtually to the President that you believe in him and in his policies, but that you intend to send a man to Washington who will assist him only grudgingly, if he assists him at all, and who more probably will do all that he can to block the very program of which you have approved in your presidential vote. The President and the representative are both constituent members of the same organized machinery for accomplishing definite legislative results. Consequently, though it may sound like fine independence to vote one's congressional and one's presidential ticket entirely without regard to each other, by doing so one is inevitably lessening the efficiency of this necessary machinery.

Again, do not understand me as making a too sweeping assertion. I do not mean that such splitting of the ballot is never justified. I mean it is only justified in unusual circumstances and after the most mature consideration. I have already said that the emergency may arise where the punishment of an individual for really intolerable conduct becomes a more vital issue than the question of a proper legislative program. The only thing I urge is that you should make this moral issue perfectly plain to your own minds.

You may protest against what I have been saying on the ground that if you vote for the right kind of Congressman he will always vote for the best measures; if you vote for the best President he will always work for the best measures, so there will always be an alignment of the best men working for the best good of the country. As a matter of fact, things do not work that way. No matter how competent a man you send, he cannot be a wholly independent agent after he takes office. For efficient legislation he must become a part of the general machinery. Furthermore, this is necessary for efficient government. If you are not convinced either of the fact or of the necessity of it, I will simply ask you to take the matter for granted for the moment in connection with what I have said above and in justification of what I have said. I shall return to this matter later in discussing the relation of the representative to his party.

So far in this general lecture on the relation of the voter to his representative, I have been discussing the question of what should determine the voter's action in the choice of representative. Leaving that now, we come to a second question, namely, what the relation of the voter should be toward his representative when he has once been elected. For the moment I shall go on the assumption that you, as voters, will expect your representative to be there not as your personal agent to attend to your personal affairs, but as a thinking legislator to act according to his best judgment for the general good. Or, if it be your fate to be the representative rather than the voter, I shall assume that you act on this same general principle.

Such an assumption as this is really not as simple as it may seem to you. But that is a question which I shall postpone to a later lecture. Granted for the moment that as you look forward to the future you assume that you will take this attitude, whether as voter or as legislator, I wish to say just a few words in closing on the question, not of how the representative shall treat his constituent, but how the constituent shall treat his representative. And what is more, it is a question not only of the attitude of the constituent toward his own representative, but of the voter at large toward representatives in general.

THE REPRESENTATIVE 85

It seems to me that one of the chief obstacles to good representative government in legislative matters is a somewhat widespread attitude on the part of many people to look on every politician with suspicion. It is not necessary here to define what we mean by politician or whether a line can be drawn between the politician and the statesman. Tom Reed's famous definition is well-known—"a statesman is a successful politician who is dead." This may sound cynical, but was the result of wit applied to long experience.

There is also a tendency to grant to those who are distant in time or in place a far greater degree of merit than to those whom one watches in daily life. When I said that there is a tendency to look on the politician with suspicion I meant by politician everybody who has entered politics as a career, whether he be the President of the United States or the representative in Congress or the alderman from the ward. More specifically I have in mind anyone holding office under the vote of a popular electorate, and I think that the statement I have made is true regarding this suspicion and that the fact is very unfortunate. It leads frequently to a lack of cordial support on the part of the public when such support is needed, and it leads also, unfortunately, to the breeding of cynicism on the part of the representatives themselves when, after conscientious and genuine effort, they find their motives misinterpreted and a casual assumption prevailing that somehow behind every one of their acts is some subtle and mischievous intention to advance some private interest at the expense of the public interest. It sometimes leads to the driving of good men out of public life, it sometimes leads to the cynical attitude that if one is getting the credit for misdoing one may as well get the profit from it, and in any case it works seriously against a sympathetic and harmonious action of the representative and the voter toward accomplishing common ends.

I believe that educated men are to a considerable extent responsible for this, and are responsible for it through a certain unconscious prejudice which they have not thoroughly analyzed. Again, I wish to urge upon you that it is just this kind of unconscious prejudice on the part of the voter, unsupported by reliable data in the way of actual facts, which I consider one of the immoral characteristics of the voting public.

It is true that there are many people who cannot understand why a man should care to go into public life with its turmoil, its jealousies and struggles, and from their own point of aloofness they are inclined to rank such men as of an inferior moral tone. On the other hand, there is another class of men, very practical, devoted solely to money getting, who cannot understand that men can have other motives than those of material profit. Many a hardheaded business man says of the politician that he can not adopt such a business "for his health." He assumes that he must have a personal and sordid motive in everything he does. Here we have two opposite extremes: one the purely material, who does not believe that anybody can have a higher motive than his own; the other the idealistic, spiritual type, who cannot recognize that some other motive than his own or some other ambition than his own can be free from a sordid taint.

The first thing for you to recognize, then (and I consider it a positive duty for you to do so), is that the desire for distinction in political life is in itself an absolutely honorable desire and may lead under the stress of emergency to almost the highest type of public service. It is frequently hard to distinguish between pure ambition, the desire for personal distinction, and the desire to perform public service. In the case of most great men in the field of politics the two have doubtless been intermingled. I have already referred to what the stress of emergency may do, and in the case of many a great man the early motive of personal ambition became, under such stress, entirely lost in the desire to serve the common good. But if it be true, as Milton has said, that ambition is "the last infirmity of noble minds," I think that few of us need be afraid of pleading guilty to it. There are so many other and meaner infirmities to which we are all subject that as the world is now constituted no one need feel a sense of shame when someone points a finger at him and charges him with ambition.

If we grant, then, that in general a career of politics is a perfectly honorable one,—and grant, furthermore, that the more honorable we believe it the more honorable it is likely to become,—the first duty of the voter is to give credit for such sense of honor to his own representative and to representatives in all cases where he has not actual evidence to the contrary. In other words, I should say that it was a distinct moral duty for you as voters to credit your representative with honesty of motive and at least a reasonable intelligence in the conduct of affairs. I do not mean to say that his conduct should not be carefully scrutinized. On the contrary, I believe that one of the greatest moral services, even if one of the most unpleasant duties, of the public-spirited citizen is to watch continuously the conduct of public officials who represent him. What I object to is that casual and cynical attitude of suspicion or even of open accusation which is so common without any basis of knowledge whatsoever. I am afraid that no body of men are probably so guilty of it as the body of college graduates who somehow arrogate to themselves an intellectual and moral superiority, when the first lesson of their education should have been that their opinions should not be swayed by unreasoning suspicion or their conversation tainted by unknowing accusation.

Now the above has been somewhat general and, as I am talking quite informally and directly to you, I shall put what I mean quite bluntly. It is this; that the representatives of the people in either branch of Congress are probably much more honest and decidedly less intelligent than you young men think them to be. There is a certain glamor about positions of this kind and I have no doubt that you exaggerate in your minds the capacity of the average Congressman and what you consider the great genius of the few leaders who have made themselves conspicuous. On the other hand, just as you make them in your minds more than plain human beings in the matter of intelligence, you make them rather less than plain human beings in the way of plotting, scheming, and planning. This is partly due to the fact that they are distant from you; partly due also, I suppose, to a certain tendency of youth to exaggerate all qualities whether good or bad. I have heard my father remark, after a long life of varied experience with all classes of men, that the longer he lived the more he came to trust in the honesty of men and the less he came to trust in their intelligence. I believe this is the conclusion that most intelligent men will come to as time goes on.

There is an interesting parallel to be drawn regarding the way in which we look at the statesmen of foreign countries. They are inclined to loom big through the haze of distance. The story is told (although I will not vouch for the truth of it) that a well-known American citizen of foreign parentage once appealed to Mr. Blaine, saying that he thought it was time for him to give up his business activities and devote his energies to the public service, and that he would like Mr. Blaine's advice as to whether he should remain in this country and run for Congress or return to England and stand for Parliament. To which Mr. Blaine is supposed to have replied: "It all depends on what kind of a reputation you want. If you want a reputation for statesmanship in England stay here and go to Congress. If you want a reputation for statesmanship in this country go back and stand for Parliament."

One who reads the comments of the foreign press upon their own statesmen and the statesmen of other countries, or who talks with men of different nations on such subjects, cannot help being amused at the way in which each gives credit to the statesmen of some foreign country for a certain diabolical cleverness. The German thinks that the foreign policy of England has been the result of the most astute, but most unscrupulous statesmanship. On the other hand, he is inclined to criticise the ministers in his own country as men without conviction or foresight, who weakly allow themselves to be trampled on by the ruthless statesmen of other countries. The Englishman, however, is likely to think of his own leaders as somewhat kindly but almost childlike men of high moral standards, who

THE REPRESENTATIVE 93

are constantly being imposed upon by foreign statesmen of great shrewdness who recognize no higher principle than that might makes right.

This comparison may seem somewhat far-fetched, but it illustrates pretty well the attitude of the average man, and especially the average young man, in his judgment of the men who have been elected to govern our affairs. The older man, perhaps from closer acquaintance, while exaggerating all the qualities of the politician from some other part of the country, is likely to take a scornful attitude toward his own representative. This, it seems to me, is in many cases utterly unfair. It is said that familiarity breeds contempt and doubtless it is true that in most cases close familiarity with any individual removes much of the glamour of what had formerly been supposed to be his great superiority. But if familiarity breeds contempt it also usually breeds affection and good feeling. It is true that when you come to know a man whom you thought was great you begin to doubt his greatness. It is also true that when you come to know well a man whom you thought was a scoundrel you begin to doubt his rascality.

I think that the politician often does not get a fair show from his constituency in this matter. They come too often to doubt his ability without coming to give him due credit for his honesty. I repeat in conclusion what I said to start with, that the duty of the voter is to recognize his representative as an ordinary human being trying to do his best, not a master of deep-laid plots and probably not a great master of intellect from whom a solution of all problems can be expected, or still less from whom he can properly expect a vote more intelligent or a stand more courageous than he could expect from himself when placed in the same position. In fact, I think the enthusiastic reformer who is frequently disgusted with his representative might be reminded of the perhaps vulgar but very human advice to the young soldier in Kipling's poem:

> When half of your bullets fly wide in the ditch,
> Don't call your Martini a cross-eyed old bitch,—
> She's human, as you are,—you treat her as sich,
> And she'll fight for the young British soldier!

There is one third and last point regarding the duty of the voter to his representative which I can make very briefly. The voter in the average district does not have the chance to vote for a great party leader. He can only vote for one of the subordinates. To be sure such a subordinate frequently starts out with dreams of immediate and brilliant success in forcing himself into the upper councils of his party through sheer force of ability and eloquence. Similarly his enthusiastic supporters expect great things of him. But if one is to reckon with facts, we must recognize that this in most cases is an impossibility. However great the disappointment to the individual young statesman may be, it is a duty of his constituent not to expect any such immediate results and not to turn against him because he fails to achieve the impossible.

It is not necessarily a sign of incapacity on his part. It is simply the inevitable result of our system of government, in which the really young man has much less show for leadership than was the case a hundred years ago or than is the case in England at the present time. It is seldom the young man of dash and brilliancy who rises to the top in our national councils. It is rather the veteran who has fought his way up from the ranks step by step. This is not due merely to the fact that committee promotions are regulated largely by seniority of membership in the House, but it is due also to the increasingly unwieldy size of the national legislature and to the fact that we do not have in this country, as is the case in England, a distinct ruling group who can advance, almost as rapidly as they please, a young man of marked promise and can make sure of his continuous election by providing a safe borough for him in case of any temporary disaster. The result is that most of our parliamentary leaders are well advanced in years. A man who can rise to the leadership of his party on the floor of the House at the age of fifty is spoken of as "a young leader."

The result is that continuance of service is of more importance than individual originality. It is practically impossible for a leader to arise out of a district which is always changing its mind regarding its representative, and yet it is often such a district which bewails the fact that its own representative seems to have so little influence in national councils. The reply is that if a constituency wishes to have a leader they must keep him there long enough to become a leader. I do not mean that leadership will come simply from long service. There must be capacity in the representative as well, but practically no degree of ability will bring him to a position of real power without continued service. The result is that it is not uncommon for the voter to turn against a representative because of his lack of influence, and by changing the representative make it impossible for that district to figure in the leadership at all.

The phrase "rotation in office" has been very popular in this country, especially in the past. It was the idea that everybody in turn should hold office, whether as a duty or a privilege. If office holding was a duty, everybody ought to take his turn. If there was something in it of advantage to the individual, everybody ought to have his share.

98 POLITICIAN, PARTY AND PEOPLE

This has been very familiar doctrine even down to modern times in the case of state legislatures, and in some states there has been almost a standing rule that a man should not serve more than one, or at most two, terms. The resulting incompetence of state legislatures hardly needs to be commented upon. The only result was that in some small districts nearly everybody could have the distinction of having once in his life been a representative at the state capitol. But this principle no longer has such a hold as formerly and nobody would advance it as a general rule in national affairs.

On the contrary, if what I have said regarding the insignificance of the average representative in Congress and the great power of the few leaders is true, it would seem to be a duty of the constituent to give a hard working, intelligent representative every chance to rise to a position of greater influence. This is something which I think the voter should carefully consider. I mean that in case of doubt he should always lean toward the incumbent for the time being. Both efficiency and power increase with the length of tenure, and this fact is a fact which the voter must carefully keep in mind in making his choice, not only in the election but in the preliminary nominations. A man may have served in Congress, let us say, three terms. A rival candidate appears for the nomination in that party and you feel that, on the whole, the new candidate is superior. What is your duty in the matter? Should you vote for the better man? In some cases you may decide that you conscientiously must do so. In other cases you may conscientiously decide that if you keep the other man in office he will ultimately become better than his rival candidate could within a given period of time. In other words, he has so much time to his credit. His efficiency has been increased by so much. The other man must begin at the beginning. In so far as you wish a man from your section to have a position of prominence, to be one of the men who really frame national policies, really control national affairs, you must be ready to stand by him as loyally as you conscientiously can.

CHAPTER IV

THE REPRESENTATIVE AND HIS CONSTITUENCY

The subject to which I wish to call your attention in this fourth lecture is the relation of the representative to his constituents. Once having been elected by their votes and dependent upon their support, what is his duty to them? This involves at the outset the whole question of whether a representative in a legislative body should be independent in thought and action, working and voting for what he considers the best interests of the nation at large, or whether, on the contrary, he is merely the agent for his particular constituency, pledged to work and vote for what may be to the particular interest of his district. This is a question as old as representative government and one which was discussed long before the United States became a nation at all. I wish, however, to point out one thing in the beginning which you should keep clearly in mind. I can do so best perhaps by quoting a sentence or two from a lecture by President Hadley entitled "Workings of our Political Machinery" (published in his "Standards of Public Morality"), a lecture to which I shall have occasion to refer several times in the remaining lectures of this course.

Mr. Hadley says: "A number of congressmen go to Washington pledged to act in the interests of those who sent them. This pledge is not an explicit one. There will always be men who disregard it in certain emergencies, and who prefer the high claims of the country to the lower claims of the party or district. But these cases will be relatively few."

What I especially call your attention to in this passage for the moment is that he groups together the "lower claims" of "party" and "district" as contrasted with the higher claims of the country at large. I think that it is important to keep the question of the claims of party and the claims of district quite distinct. In fact, I shall try to prove to you later that one of the most effective causes leading in the last few years to a relative lessening of the demands of particular districts is the increased necessity of strict party loyalty. They may both be "lower claims," but one, I think, tends partly to eliminate the other. This, however, is a matter for later discussion. I speak of it here so that you may keep your minds clearly on the fact that it is the question of service to the district rather than of service to the country at large which I am now discussing.

I do not know that I can say anything new upon this subject. I should like to give you the views of many different thinkers of different types, but our time will not permit. You young men, probably with scarcely an exception, take it for granted that the higher ethical duty is service to the country as a whole, and I certainly agree with you. On the other hand, you are probably deeply shocked at the very suggestion that it is the duty of a Congressman to act simply as the agent of his own constituents and fight solely for their interests, while I, though disagreeing with this theory, am not shocked by it at all. I know very able men who defend it upon grounds which we may hold to be untenable, but which are in no sense immoral, or which do not even show a lower moral conception than that held by those who disagree with them.

I have found in my reading that on most political questions Edmund Burke always speaks more wisely than anyone else, as well as more eloquently, and a perusal of some of his speeches shows how little the ethical problems of politics have changed in a century and a half. His is probably the classical expression of the theory that a parliament is not, as he puts it, "a congress of ambassadors from different and hostile interests," but rather is a body representing one nation with one interest. I wish to quote somewhat at length from his noble speech "To the Electors of Bristol," delivered just after his election to Parliament from that city in November, 1774. He says:

"Certainly, gentlemen, it ought to be the happiness and glory of a representative, to live in the strictest union, the closest correspondence, and the most unreserved communication with his constituents. Their wishes ought to have great weight with him; their opinion high respect; their business unremitted attention. It is his duty to sacrifice his repose, his pleasures, his satisfactions, to theirs; and, above all, ever, and in all cases, to prefer their interest to his own. But, his unbiassed opinion, his mature judgment, his enlightened conscience, he ought not to sacrifice to you; to any man, or to any set of men living. These he does not derive from your pleasure; no, nor from the law and the constitution. They are a trust from Providence, for the abuse of which he is deeply answerable. Your representative owes you, not his industry only, but his judgment; and he betrays, instead of serving you, if he sacrifices it to your opinion.

"My worthy colleague says, his will ought to be subservient to yours. If that be all, the thing is innocent. If government were a matter of will upon any side, yours, without question, ought to be superior. But government and legislation are matters of reason and judgment,

THE CONSTITUENCY

and not of inclination; and, what sort of reason is that, in which the determination precedes the discussion; in which one set of men deliberate, and another decide; and where those who form the conclusion are perhaps three hundred miles distant from those who hear the argument?

"To deliver an opinion, is the right of all men; that of constituents is a weighty and respectable opinion, which a representative ought always to rejoice to hear; and which he ought always most seriously to consider. But *authoritative* instructions; *mandates* issued, which the member is bound blindly and implicitly to obey, to vote, and to argue for, though contrary to the clearest conviction of his judgment and conscience; these are things utterly unknown to the laws of this land, and which arise from a fundamental mistake of the whole order and tenour of our constitution.

"Parliament is not a *congress* of ambassadors from different and hostile interests; which interests each must maintain, as an agent and advocate, against other agents and advocates; but parliament is a *deliberative* assembly of *one* nation, with *one* interest, that of the whole; where, not local purposes, not local prejudices ought to guide, but the general good, resulting from the general reason of the whole. You chuse a member indeed; but when you have chosen him, he is not a member of Bristol, but he is a member of *parliament.*"

I do not believe that anything can be added to this eloquent statement of Burke's in favor of the independence of a legislative representative in exercising his own best judgment and following his own conscience in working for the general good. There may be some statement of the case on the other side in literature, replying to this classical argument of the greatest of English political philosophers, but if so I have never seen it. I mean a statement by a conscientious believer in the doctrine that the representative should be merely the agent of his constituents. President Hadley, who does not believe in this theory, has made an interesting statement of the position in that essay

THE CONSTITUENCY

to which I referred at the beginning of this lecture. He puts it as follows:

"Many men who admit in theory that their duty to the country is greater and more important than their duty to their constituents disclaim their responsibility for putting this theory in practice. They say frankly that while our government would be a better one if everybody recognized that principle, it will only introduce confusion and injustice today if a few good people work for the benefit of the nation while a great many people who are not so good have only the claims of the party or the district in view. They hold that the selfishness of a number of sections of the country, each pulling in its own way, will produce a fairly salutary general result for the country as a whole. Equity between the different parts becomes in their minds a more prominent consideration than the general interests or safety of the whole, which they are willing to trust to Providence to take care of. They are in the mental attitude of the little girl who saw a picture of Daniel in the lion's den, and whose sympathies were excited, not so much by the danger or probable fate of the prophet, as by the disadvantageous position of a little lion in the corner who, as she said, probably wouldn't get anything."

This statement of President Hadley's is admirable, but I do not think it tells the whole story. You see, he gives it as an argument by men who defend such a practice as a necessary matter of expediency under given conditions, but who, as he says, frankly believe that our government would be a better one if everybody recognized the other principle. This would imply that no one conscientiously believes that even in principle the best government can be secured by averaging the conflicting interests of particular localities. I know of men of no mean ability and of long political experience who, however, do maintain this view. In doing so they are carrying their ideas of individualism and democracy to a strictly logical conclusion.

From your study of economics you are familiar with the fact that the great school of political economy, which was founded by Adam Smith and dominated English thought through the first half of the nineteenth century, started on the assumption that each man knows his own interest best and also knows best how to get it: that consequently the interests of society will be best served by allowing complete liberty to the individual to follow his own interest with only such restrictions as will protect the rights of others. Adam Smith said that through the interplay of these rival forces of self-interest men are led "by an invisible hand" to best serve society. He comes, then, to the conclusion that the interference of the legislator in commercial matters at least is "as impertinent as it is harmful."

Carried to its logical conclusion, such a doctrine of individualism would seem to warrant the elimination altogether of the effort to work for the general good. If each individual really knows what is best for him, he will work to secure that end. Thus, within any given district, each man voting intelligently for his own interests, the expression of the majority will inevitably be an expression of what is best for that community. Then let each district be represented in the national council and let each representative work solely for the interest of his particular district, and equally inevitably the result of majority action must mean the adoption of such legislation as is for the best interests of the community as a whole. I once heard this theory very forcibly stated by Thomas B. Reed of Maine, one of the greatest speakers the House of Representatives has ever had.

A friend of mine, who is well known in the political arena and who has given much thought to this question, believes that the proposition that this interplay of individual interests will bring the best general result is not really a theory at all but rather an exact mathematical demonstration. He cannot see how anyone can dispute it. To him it is as simple as an equation in algebra or as the proposition that the resultant of two forces working at right angles is motion along the line of the diagonal. Unfortunately it is not necessarily true, however, that motion along the line of the diagonal is

THE CONSTITUENCY

for the best public good. Human affairs are not determined by mathematical principles. For instance, a very valid objection to arbitration of industrial disputes is that too often the arbitrators do not really arbitrate according to some definite principle but merely "split the difference." How harmful such a practice may be was shown by King Solomon when he suggested an equal division of the child between the two rival claimants.

In any case this theory involves two premises; first, that every man does know what is best for him and, secondly, that he knows best how to get it. These are exactly the premises which my friend accepts. He is consequently in favor of every movement toward making legislation by the people as direct as possible. The people, he believes, cannot go wrong when no restraint is put upon their action in seeking their own ends. They have the right to what they want and they are only kept from securing this right by constitutional and political limitations to their power. He, therefore, favors not only direct primaries for every elective office, even that of President, but the initiative, the referendum, and, of course, the recall. Under such a theory the very object of the recall is to force the representative to vote on every measure exactly as the majority of his constituents want him to vote. If he does not do so in any particular case he destroys, you see, that beautiful mathematical equation.

I do not wish here to enter into any discussion of these new proposals to give a more direct and rapid expression of the will of the majority. We should note, however, that they are bound to make for a political system under which the member of Congress is no longer a man of independent judgment, but simply an agent to express the desires of the particular group which he represents. The extraordinary thing is that we frequently find one and the same man advocating these measures and ardently urging every form of direct legislation and at the same time condemning Congressmen for their subserviency to "mere popular whim." It is, of course, the old, age-long problem of direct government by the people versus representative government. Whichever attitude you may take on this question, you should at least keep clearly in mind all the consequences involved. If you wish to have great statesmen of courage and independence, whose judgments guide the policies of the nation, you must favor some form of truly representative government, even if it carries some evils with it. On the other hand, if you wish the people of each district to have the opportunity to give immediate expression to their desires, you must not expect to have leaders of this character. I believe this at least to be true in a system such as ours where a Congressman practically always represents the district in which he lives and must at least be elected from his own state.

I do not know how far this "agency theory" may appeal to any of you. I have already said that I subscribe rather to the idea expressed in the noble words of Burke. I will assume, then, for the rest of our discussion of this subject that you do the same. In any case, I should like to suggest again that I do not believe that the "agency theory" is carried out even in practice today as much as it was a dozen years ago, or as much as many people believe it to be, and I believe that one reason for this lies in the control of the individual representative by his loyalty to party or by the pressure of the party caucus. This is a topic to be discussed in the next lecture. There still remain intricate problems regarding the duty of the representative to his constituency, even if we agree that in matters of general legislation he should be a free agent following his own judgment and conscience.

After all, it must of course be remembered that, whatever theory we may hold regarding the relation of the representative toward public policies in the matter of independence and freedom of judgment, he is really a *representative;* that is, he represents the particular district from which he is elected and the men who vote for him have not done so solely from the idea that he should be a great statesman exercising his mind all the time on the problems of national welfare. They want part of his mind and part of his time themselves and, what is more, I think they have a right to expect a certain amount

THE CONSTITUENCY 115

of attention from him. It is sometimes possible for a man practically to disregard his constituency and tell them that he will pay no attention whatsoever to their demands in the matter of patronage or appropriations, or their requests for assistance in personal matters, however legitimate. Such men, however, if they are to keep their position in Congress at all, must have already achieved such a commanding position that their districts take sufficient pride in the power of their representative to offset their dissatisfaction at the neglect of their interests. Mr. Reed, for example, who expressed the agency idea theoretically, was powerful enough to disregard the importunities of his constituents in practice.

The average Congressman must recognize the fact that he is expected to attend to a great many matters on behalf of the people who have elected him. Here is where his ethical problems are likely to become acute, but I beg of you at least that you will realize that, like other problems I have suggested to you, they are not problems of the present time alone. It is very easy to make assertions regarding "the corruption of modern politics" or the lowness of "modern moral standards" and to hark back to an earlier day when great men lived who had no thought save for the welfare of the public. But it does not take much reading in the letters and diaries of, say, the eighteenth century to realize that the mixture of selfish and patriotic motives was as prominent then as now. Even great philosophers campaigned for places of emolument either at the universities or in public service in a way which would seem beneath the dignity of even the youngest instructor in these days. Certainly the requirements made by his constituents on a parliamentary representative and the degree to which it was necessary for him to attend to manifold personal interests were not only as great, but were probably greater a century and more ago than they are today. I have already quoted at some length from Burke's speech "To the Electors of Bristol" of 1774. Six years after, in 1780, he again spoke at the Guildhall in Bristol in a speech entitled "Upon Certain Points Relative to His Parliamentary Con-

THE CONSTITUENCY

duct." He felt it necessary to defend himself against four charges of which one was the neglect of his constituents, and on this point he speaks as follows:

"With regard to the first charge, my friends have spoken to me of it in the style of amicable expostulation; not so much blaming the thing, as lamenting the effects. Others, less partial to me, were less kind in assigning the motives. I admit, there is a decorum and propriety in a member of parliament's paying a respectful court to his constituents. If I were conscious to myself that pleasure or dissipation, or low unworthy occupations, had detained me from personal attendance on you, I would readily admit my fault, and quietly submit to the penalty. But, gentlemen, I live at an hundred miles distance from Bristol; and at the end of a session I come to my own house, fatigued in body and in mind, to a little repose, and to a very little attention to my family and my private concerns. A visit to Bristol is always a sort of canvass; else it will do more harm than good. To pass from the toils of a session to the toils of a canvass, is the furthest thing in the world from repose. I could hardly serve you *as I have done,* and court you too. Most of you have heard, that I do not very remarkably spare myself in *publick* business; and in the *private* business of my constituents I have done very near as much as those who have nothing else to do. My canvass of you was not on the change, nor in the county meetings, nor in the clubs of this city. It was in the house of commons; it was at the custom-house; it was at the council; it was at the treasury; it was at the admiralty. I canvassed you through your affairs, and not your persons. I was not only your representative as a body; I was the agent, the solicitor of individuals; I ran about wherever your affairs could call me; and in acting for you I often appeared rather as a ship-broker, than as a member of parliament. There was nothing too laborious, or too low for me to undertake. The meanness of the business was raised by the dignity of the object."

You see, then, that in Burke's day the representative was pestered by a great many local interests and that even so great a man as Burke was obliged to stoop to what he himself called "ship-broker's work." Of course, it really is the duty of the Congressman to take care of the interests of his constituents in every honorable way. Although we do not accept the "agency theory" in the field of measures of public policy, the Congressman must to a certain extent be the agent of the members of his district and assist them in matters where they have just cause of complaint or just claims.

For instance, besides the public acts passed by Congress, there is a large amount of what is known as private legislation; that is, legislation affecting the position only of some individual. As good an illustration as any of this class of legislation are the private pension bills with which every Congressman has to deal. There is a general pension law describing the general rules under which pensions will be granted. It is quite possible that under these general rules pensions may be granted to quite undeserving cases.

On the other hand, it is equally possible that, through some technicality, very deserving cases may be excluded under the general rules. In such a case an appeal is made to the Congressman and a private act granting relief to that particular person may be introduced. Naturally, the Congressman who gets a pension for anybody makes himself popular with that person and his or her friends. He runs little danger because no considerable number of people are likely to vote against him for having secured pensions in this way. The result is that his own interest would usually lead him to try to get bills through, not only in cases which he really thought to be deserving and where, under the real spirit of the law, a pension should be granted, but also in cases where he knows that there are no just grounds for the claim.

It is just here that the test of a Congressman's conscientious devotion to public service comes in. It is his duty toward his constituents to do what he can to see that their just claims are recognized. It is his duty toward the country at large to see that

THE CONSTITUENCY

none but just claims are granted. In some cases it may be very difficult for him to decide. The brave Congressman will stand firmly against improper claims, but he will get little credit for it. Unfortunately, where bravery brings no rewards it is much easier to be accommodating than to be brave.

Another line of activity in which the Congressman is bound to engage is concerned with the administration of federal affairs in his district. There may be incompetence in the post office or in the custom house. There may be rules of a department which work hardship in the case of his particular district. One may say that the Congressman should not bother himself with matters of this kind; that the dissatisfied party should appeal directly to the administrative department concerned in order to secure any mitigation of the evil. But even the best departments are necessarily bureaucratic and likely to be somewhat scornful of local objections. I believe it is the duty of the Congressman in such a case to devote his time to the interests of his constituents in seeing that these matters are fairly considered by the administrative officials. After all, a Congressman will be listened to with much more attention than some simple constituent.

This fact unfortunately, however, leads many Congressmen to attempt to use their position of influence to browbeat honest administrative officials into acting contrary to the public service for the sake of their particular district, or, what is worse, for the sake of some particular constituent. It is nothing short of disheartening to note cases which too frequently arise of Congressmen actually making threats to block appropriations, or to somehow hamper the administration of a particular bureau, unless the official yields a point in favor of his particular claim.

This is one of the problems which ought to be fairly simple in the mind of a conscientious representative. Where it is a matter of really improving the administration of federal affairs in his district it is one of his duties to the district to use his influence. No one, however, could question the immorality of his yielding to the demands of certain interests in his district to urge increased laxity of administration or to impede an honest official in his efforts to do his duty.

The attitude of some people in this regard is so naïve as to indicate that they are not so much immoral as unmoral. I knew of one case where a man appeared before an official of the treasury regarding a customs matter and seriously urged that certain action on his part should be allowed. When it was pointed out to him that this was strictly contrary to law he naïvely said, "Yes, but when the bill was up I told Senator Blank how that clause in the law would work against me and he said that it was impossible to make any change in it, and all I could do would be to find some loophole in the law." In this particular case there was no malign intention of corrupting the official, but such a frank confession to the very person whose duty it was to administer the law shows an extraordinary attitude toward the problems of public duty.

One of the most difficult problems is the problem of patronage. Here again, although there is much that is vicious in our system, you should recognize that it is entirely proper for the representative to have his say regarding office holders in his own district. The Congressman is frequently looked upon with suspicion every time he asks for certain appointments to be made, as if in some way it were a dishonorable thing for him to take part in such matters. The head of an administrative department may adopt the attitude that it is solely his business to make appointments and that he will make them solely for the good of the public service. Of course, this is the principle on which appointments should be made. At the same time the Congressman is entitled to an opinion as to which men in his district are best suited for any particular positions. The difficulty arises where the Congressman uses his influence to have men appointed who will be useful to him personally. It seems to be part of our whole machinery of government for offices to be largely awarded as a reward for personal and party service. Much, of course, has been done to eliminate the evils of the patronage system through the adoption of the civil service system. But there still remain a good many offices appointment to which is a personal matter, and where the Congressman finds it very hard to eliminate the consideration of how the appointment is going to affect him individually.

Here again the difficulty lies in knowing what is the honest and right thing to do under given circumstances. A man may honestly feel that his continuance in Congress is a desirable thing for his community; that he can both represent the interests of his district more efficiently than a rival candidate and that he can also serve the country better; but nominations and elections depend very largely upon a man's standing with the political organization of the community and something more is needed to maintain the tenure of one's position than conscientious work in the interests of the country at large. Most men at least must pay some attention to the effect of their influence in keeping themselves "solid" with the organization. It is very easy for a man to delude himself into believing that such appointments as are likely to strengthen his position are the appointments which are best for the public service. It is here that any of you who go into this career will find the greatest strain put upon your consciences and will be most likely to fall below the high standard which men of your training and education should uphold.

Another matter which as a Congressman you will find occupying much of your attention and again making a severe test of your moral fibre is in the matter of getting appropriations for public works within your district. Some men in Congress maintain their positions almost entirely because of the success they have shown in always "looking out for the district." Each section of the country selfishly wishes to get as much money as possible out of the national treasury and the so-called "pork barrel" bills are those which are surrounded by the most unsavory methods of log-rolling and trading. Here again the test is a severe one because a Congressman may well feel that, if a certain scale of expenditure on the part of the national government is to be adopted anyway, it is really his duty to see that his community gets a fair share. It is the case of the little girl of Mr. Hadley's story who feared the little lion would not get his fair share of Daniel. But this desire to be sure that his district gets its share is just that which makes it so difficult for Congress as a whole to maintain a policy of scrupulous economy, or even honesty, in matters of this kind.

The way in which government contracts have been secured for the dredging of certain rivers, or the establishment of some government institution, or the awarding of contracts for public buildings, shows the extent to which our representatives have fallen below that standard of moral duty which would be expected of them by any young man starting into public life with an enthusiasm for the right. In any individual case it may be very difficult to draw the line. It is obviously quite proper for the representative to use his influence to secure the building of proper and suitable federal buildings in the towns of his district. On the other hand, it is obviously immoral for him to attempt to secure an appropriation in the way of some government work such as dredging a river when he knows that no possible economic benefit at all commensurate with the amount of the expenditure can be derived; that is, when it is simply pouring the funds of the federal treasury temporarily into a district so that the contractors, the traders, and the laborers profit at government expense without leaving any permanent gain as a result of such expenditure. But between the obviously proper and the obviously wrong thing there are many instances where the issue is a very grave one and where the individual, if he aims to live up to the moral standard with which he started, must search his mind and heart with perfect frankness to determine which line of action he should take.

One of the most flagrant cases of dereliction of duty on the part of Congressmen is in failing to support the efforts of an administrative department toward a more economical management of its affairs. Many useless offices are maintained in order to give more jobs to the members of their districts. There are useless army posts, navy yards, and custom houses. It is, of course, quite possible that a Congressman should honestly differ with the administrative department regarding, let us say, the maintenance of a custom house at a particular port. He might feel that the commercial interests of his section would really be jeopardized by a too radical effort at economy. In such a case he may honestly urge the claims of his district as vigorously as he likes. It is much more likely to be the case, however, that what he fears is that some good supporter of his own will be put out of a job and that his influence in the district will be diminished by this administrative improvement.

One of the most amusing things to watch in our politics—at least in a cynical sense amusing—is the continuous criticism of the administrative departments by Congress for their extravagance and the continuous blocking of many honest efforts at economy by these same Congressmen. Most Congressmen believe in economy in general, but it takes an unusual one to believe that the federal government should economize in his district. I think it is only fair to ask you to remember, when you read speeches attacking the extravagance of the federal government, that such extravagance is frequently forced upon an unwilling department, even after repeated appeals by it for reform, by a Congress really more eager to maintain their constituents in offices than to protect the public treasury.

Still more difficult, perhaps, than matters of patronage and federal expenditures is the question of the representative's attitude toward the business interests of his particular district. If we adopt the "agency theory" the matter, of course, becomes relatively simple since, whatever his own views may be regarding what policy is for the welfare of the country as a whole, he would advocate such policies as were for the interests of his community. If, however, we believe that the Congressman should vote according to the dictates of his conscience in such a way as to serve the interests of the whole nation, it would seem that he should pay no attention whatsoever to the special effect upon his own district of legislation which he believes to be for the common good.

THE CONSTITUENCY 131

I think you will all agree that this is the higher attitude and that if he cannot hold his position by following a broad, patriotic policy of this kind he must simply make his defense before his people on a higher moral plane and leave them, in case they wish to pursue a purely selfish policy, to send a representative who is either more subservient or who conscientiously believes in some different line of public policy. But when you are practically in a position of this kind the problem is not quite so easy as it seems in the lecture room.

One of the most important lines of public policy which affects greatly and in varied manner the business interests of different sections is the tariff. What is the duty of the Congressman who believes in a large reduction of the tariff and the adoption of the principle of "tariff for revenue only" in the case of industries in his own community which he thinks would be injuriously affected, or which might even be forced out of existence altogether? The obvious answer is that he should courageously take his stand according to his conscience on the general policy and not attempt to make any exception in the case of some particular local industry. But suppose that he knows that the measure which is to be adopted practically is going to be a compromise matter and is not going to carry out any theoretical principle; that it will carry many rates which have been the result of the special care taken by other representatives for the industries of their districts. If he is willing to be a lone hero who makes no compromise whatsoever, how about his feeling of responsibility toward those who have elected him? Shall his district be made the lone victim? If there were some absolutely clear-cut principle of tariff making, and every representative would vote for or against the measure according to some such principle, the problem would be clear enough. But tariff acts are a mass of actual rates and these rates are matters of compromise and adjustment.

I think I can see a certain ground under such conditions, since the interests of other districts are being carefully watched and efforts being made to protect them, for a man's claiming that he ought to go a certain way along the same path in looking out for the material welfare of his own neighbors and supporters; that in fact he is derelict to his duty to them if he does not. I do not say that I believe this is sound. I say only that I can understand a man's having here a genuine moral problem. Again it becomes so much a matter of degree. It is not merely the question of whether or not he shall try to secure every possible advantage for his district. It may be the question whether he ought not to do something simply to give his district a fair show with the others so that any sacrifice that is to be made under the new policy will be a sacrifice fairly and evenly distributed, and so that it will not bear with extreme and unjust force on his constituents. The trouble with such an attitude is that it does very largely take the principle out of the matter altogether, and makes the problem of each separate industry, and the amount of duty on its products, an individual problem where his conscience may be easily stilled and his moral fibre weakened with each successive concession.

134 POLITICIAN, PARTY AND PEOPLE

As I have already suggested, one thing which helps the representative out in matters of this kind is the force of the party organization and the party caucus. It is interesting to see the way in which caucus action has to some extent changed the problem of loyalty to district. It has substituted in certain measure the idea of strict loyalty to party and that is the problem which will concern us in the next lecture.

CHAPTER V

THE REPRESENTATIVE AND HIS PARTY

The last problem we have to consider is the relation of the representative to his own party. I will begin by referring to certain statements by President Hadley in that lecture on the "Workings of our Political Machinery" to which I have referred before and which all of you should certainly read. It is full of the wisest comment. Mr. Hadley speaks of the difficulty, under our present system, of getting efficient legislation, due to the fact that to a very large extent our representatives are not sent to Congress to make laws or to govern the country; that under our constitutional system the President cannot govern alone and Congress cannot govern alone; that this separation leads often to such a dead-lock that the representative is much more concerned with problems of place and patronage and the wants of his district—questions which I discussed briefly in the last lecture—than he is with questions relating to legislation in behalf of the general welfare. Mr. Hadley thinks that as a necessary consequence the political boss has become a more powerful figure in actual government than the elective representative of the people. I cannot consider in detail the many interesting suggestions which he makes in connection with these matters and I agree with him very largely in all that he says when his remarks are applied, as he suggests in one passage, to the workings of our political machinery at the end of the nineteenth century.

What I wish to suggest here is that I believe we have been going through a change in recent years which is of the utmost importance and which the future historian may write down as revolutionary in character. In his preface Mr. Hadley suggests that if anyone should take up the book a few years later he hopes that, though the events in the foreground may have changed, the reader will find the underlying principles yet of value. This was written in 1907 and is a striking illustration of how rapidly changes may take place, or at least how rapidly we may become conscious of such changes. Unless I am completely mistaken in my diagnosis, this new development had only begun a few years before 1907 and has only come to show its full importance in the years since then.

Perhaps I can best indicate what I mean by this change by telling of a conversation I had with a bright young German who came to me on his travels with a letter of introduction about 1903. The first question he asked me was: "Who rules your country?" I began to reply by some explanation of our system of government, to which he said impatiently: "But I don't want any of your theories. I know your constitution by heart and have read my Bryce and all the other books thoroughly. I want to know the names of the men. Is it John Smith or William Jones, or who is it?" For the moment I was obliged to hesitate. I told him that if he had asked me that question a few years earlier I would have given him the names of a small group of Republican senators and I named as those who I thought could have been fairly considered the "big five" of the old days—Senator Aldrich of Rhode Island, Senator Hale of Maine, Senator Allison of Iowa, Senator Platt of Connecticut, Senator Spooner of Wisconsin, with a choice between three or four others for a possible sixth place, and suggested that to these should certainly be added the Speaker of the House and possibly one or two chairmen of leading House committees. These certainly were the men who determined more than anyone else what legislation should go through Congress, or perhaps it would be better to say of that particular group that they were the men who determined what legislation should not go through Congress. "Well," he said, "if they don't rule the country now, who does rule it?" To this I replied, "The issue is at the moment not entirely settled, but, if I am not mistaken, the country is ruled by Theodore Roosevelt."

I tell this story not merely to suggest that at that time there had been a change in personalities, but to call your attention to the significance of a great change in principle; namely, the increasing power of the President as a leader of a party with a definite program of legislation in which he takes the initiative. I think that this movement has been going on steadily ever since and, what is more, I am inclined to believe that it is an inevitable tendency and one which meets the desires of the American people. If this is so, it is no longer as true as formerly that neither the President nor Congress can govern the country, and it becomes quite possible that a strong executive, acting as a party leader and working in harmony with a group of Congressional leaders, can in the future fill the position formerly occupied by a different group of political bosses.

Referring once more from Mr. Hadley's essay, he makes a very interesting parallel between American politics at the end of the nineteenth century and English politics at the end of the eighteenth century. The passage is so significant that I wish to quote it in full. He says:

"There has been one other country and one other age in which political parties have had the same character that they have in the United States today. That was in England during the eighteenth century. And it is a noticeable fact that the English government in the eighteenth century had this characteristic in common with the American government in the nineteenth; that the executive and legislative branches of the government were so far separated that no means of harmonizing their action was provided or allowed by the Constitution. Under such circumstances the English parties at the beginning of the eighteenth century, like the American parties at the end of the nineteenth, were primarily occupied with keeping certain men in office, and the passage of legislative measures formed only a very incidental element in their plans. With this striking parallel in view, we may well believe that the separation of powers between the different departments of the government, and the perpetual threat of a deadlock thereby produced, have as an almost necessary consequence the dominion of the party manager: that Walpole and Tweed were but different specimens of the same genus; and that their power, however widely different in its methods of exercise, was an outgrowth of the same cause."

What I am suggesting here is that, just as in English politics a system has been worked out to avoid the extreme separation of powers between the different departments of government, something of the same kind is now being worked out in this country, perhaps in a somewhat blundering way, but nevertheless in a way that is going profoundly to affect American politics in the future. Under the working out of this new system, we may possibly predict for our politics in the twentieth century as distinguished from the close of the nineteenth that it will no longer be true that parties are primarily occupied with keeping certain men in office, or that the passage of legislative measures is only an incidental element in their plans.

This, I think, is a result of several factors of which I will mention three. First, the actual breakdown of the old system as a practical working force for governmental purposes; secondly, the growth of a new spirit of earnestness in our politics, due, I think, largely to the rising generation of the Middle West; and, third, the continuous increase in the size of the House of Representatives which has made that body entirely unwieldy except under some new form of party organization and party leadership. In England the difficulties arising from separation of powers between the executive and the legislative were overcome through the gradual development of a system of responsible cabinet government and the growth of the cabinet as the real executive authority. In this country no such system could be adopted without most radical constitutional changes and, although I sympathize largely with those who advocate responsible cabinet government as the best form in a democratic community, I do not believe that it could be arbitrarily substituted for the American system. There will probably be some natural evolution which will, however, bring about similar results and I find the first step in this direction in the increasing initiative of the President in legislative matters.

THE PARTY 143

Some people have been inclined to quote with a somewhat cynical smile President Roosevelt's continuous reference to "my policies," but in that very phrase I find something much more profound than the self-assurance of any individual. I will not attempt here to give any opinion as to how far it took the extraordinary personality of Mr. Roosevelt to make this conception a vital part of American political life or how far it was something inevitable which had to come in any case. The main point is that, apparently, it has come. The people are not offended by any talk about "my policies" because they now expect the President to have policies. Before President Taft came into office in 1909 he issued a formal statement of his policies, covering, as I recall it now, thirteen specific heads in the nature of legislation. I believe that this will continue to be the case in the future; that as we speak of Roosevelt policies or Taft policies, so we will speak of a definite legislative program by the name of future Presidents.

This may seem very simple and natural to you young men who have become accustomed to it, but it is a much more important change than you probably recognize. It has become necessary simply from the breakdown of the old system. What the people wanted and what the country needed was something in which to believe. It is more or less true, I think, that the old parties fifteen or twenty years ago stood in the minds of the voters for little more than the question of who should get in or who should stay out. It was essential that there should be a more definite conception of a party program and a more responsible party leadership for carrying that program into effect. By responsible leadership I mean here concrete leaders whom the people could hold responsible for carrying out the policies of their choice and whom they could reward or punish according to the way in which this work was accomplished.

I do not mean to say that we have come to the point or shall come to the point of purely one man power. I mean that we seem to be overcoming some of the old difficulties by the growth of the President as a leader in legislative matters through the power of his personality and his influence on Congress. With this, however, must go, of course, a clear-cut party leadership in Congress itself, combined with at least a working degree of harmony between the President and these Congressional leaders. Note one thing, please, in this connection, to which I can refer only briefly. If what I have said before regarding the certain tenacity of party organization is true, this new movement inevitably gives to the President somewhat more of a partisan character than many idealists want him to have. It is frequently said that when once elected to that office a man should forget his party and be simply "President of all the people." This is one of those phrases which appeal to our ideals, but which are too often used without any analysis of their real meaning. The position of President of the United States is perhaps unique among all political positions of the world. We expect him to be something more than a party leader and we also expect him to be something more than a figurehead, and this new movement especially expects him to be a leader in a legislative program.

146 POLITICIAN, PARTY AND PEOPLE

In England they have the King for nonpartisan purposes and the prime minister to carry out the will of the people at any given time regarding policies to be enacted. In this country we are coming to expect the President to be both, but, if I am correct in saying that legislative programs must first be party programs which are presented to the people on election day, it follows that the President, if he is to lead in the carrying through of such a program, must become more and more the real leader of his party. You must recognize, then, that you will be entirely inconsistent if you expect him to perform this function and yet be solely "the President of all the people." He may still remain the President of all the people in the sense that he is elected by the majority to carry through the program which they desire. If this seems to you to in some ways reduce the high dignity of the office, you should remember that efficiency in government is more important than ceremonial form and such a President can still be wholeheartedly interested in what he considers the welfare of the nation at large. If he is primarily allied with a certain party, he is at least free from sectional control, and becomes the representative of the whole country.

I have suggested that the second factor working in this direction was the new spirit of earnestness in our political life. I think it is true that a new element has grown up within the ranks of both parties which has shown increasing power in party councils and at the polls and which has been largely responsible for the overthrow of many of the old leaders. One reason why the old leaders have so completely failed to realize the importance of this new movement, and have largely lost their leadership as a result, is that they have not been able fully to recognize the fact that people are taking political problems seriously. By this I mean not simply the problem of who is going to be elected, but what the policy of the country shall be on a large number of matters of the utmost importance—the tariff, currency, conservation, trusts, and many others.

In the early days of the Republic these matters were taken seriously. In the period of great prosperity following the Civil War, when it was almost certain that one party would stay in power for a long time, the mass of people took little seriously except their business. The play of political parties seemed to them largely a game, or a scramble for spoils. Under such conditions the old character of party government was possible. Now a third period has come in which people are thinking on these subjects and are feeling deeply regarding them. Frequently they are very ignorant and frequently they are misled, but at least they are in earnest. They really expect their votes to count for something in the way of a legislative program. They even take party platforms seriously and propose to hold a party and its leaders responsible for its success or failure in meeting the obligations of its platform.

The third factor, as I have suggested, is the increasing unwieldiness of the House of Representatives due to the increase in numbers. Or at least I think this should be added to the other two factors as explaining why it is that a new form of party responsibility is being developed. In the early days of Congress the numbers were not so large but that men could independently hold personal views on many matters of public policy and could thrash out many of these questions in actual debate. The theory of our government, of course, is that these questions should be debated fully and that through the mutual persuasion of arguments an agreement could be reached which would represent the careful and intelligent judgment of the legislative body. Many people believe that this still ought to be the case and that somehow the old practice can be restored. I myself believe it is time to recognize that we cannot return to this earlier ideal. The problems of today are much more numerous and much more complex than formerly. On the other hand, the House has become much larger and the opportunity for the individual to be heard on many subjects is inevitably less. It is practically impossible at the present time, if there is to be any legislation at all, or anything approaching efficient government, to allow every representative to give voice to his own views on every subject. In many cases, in fact, the leaders must assume a very arbitrary attitude toward the rank and file. Even a man of great ability and conscientious in his study of these questions cannot be allowed to take up the time of the legislative body indefinitely. Still less can the many men who wish to talk simply to impress their constituents with their own importance be given such liberties.

To this is due one of the changes which I have already suggested in earlier lectures; namely, that under the present system the representative no longer has as much of an opportunity to stand out for the particular interests of his district as formerly. There is a party program to go through. It has been framed probably by a few leaders in consultation with the President. It represents the program of the party for the whole country and the particular claims of this district or that district can no longer be given much of a hearing. Thus in large measure party loyalty has supplanted loyalty to one's section. This is not always in the nature of a very willing loyalty and it means that members are whipped into line by those in control to stand for a national program represented by a national platform which was presented to all the people and on which they were elected.

Many of them, doubtless, when whipped into line in this manner, sympathize with Disraeli's feeling toward Sir Robert Peel at the time that the former was a young and somewhat obstreperous member of Peel's party. You may remember that Peel had been very active in working for the abolition of slavery in the British colonies. One day in his absence from the House of Commons a "snap" motion regarding the sugar duty was carried contrary to Peel's general policy by the votes of some of the high protectionists in his own party, including Disraeli. Sir Robert promptly reappeared in the House, coerced his recalcitrant members, and insisted on the vote being rescinded. This was done, but not before Disraeli had had time to rise in his seat and remark, "The right honorable gentleman seems to be opposed to slavery in every part of the world except in the benches behind him."

The chief agency for the efficient carrying

through of a party program is the party caucus. On all matters of vital policy the majority holds a caucus in which the party as a body agrees to stand by a certain measure, not only in general, but in detail. This measure, which has been framed by a committee, or more commonly by one or two leaders representing the committee (in consultation with other party leaders inside and outside the House), may, of course, be amended in caucus, although it is becoming increasingly the case that even caucus amendments are not many and that the bill of the leaders is accepted. However, in some vital cases radical action may be taken. In such a case the leaders would stand loyally by the amendments. It may even happen that the caucus will turn against the leaders altogether and entirely reverse a proposed policy. The main point is that all members of the caucus, except in rare instances (and usually they give notice at the time), are bound to stand by the measure *in toto* when it comes on the floor of the House. There may be occasional exceptions, but the general principle of the party caucus is that all amendments to be made by the majority will be made first in caucus and, secondly, that all amendments by the minority party made when the bill is brought in for the vote will be regularly voted down. The caucus measure becomes the established party measure. Men who voted for an amendment in caucus will vote against the same amendment when introduced by the opposition on the floor of the House. In the same way, of course, the minority may hold a caucus and agree to some definite bill as a substitute or some definite line of policy in opposition to the majority measure.

What shall we say of this method of legislation and the problems which it presents to the representative as to his choice between independent judgment and loyalty to the party? I confess that at one time I felt that the growth of the power of the caucus was extremely unfortunate and something to be fought in every way. I am, however, far from sure that I would take such a position at the present time. In fact, I am inclined to think that under present conditions it is about the only possible method of efficient

154 POLITICIAN, PARTY AND PEOPLE

legislation under our form of government. Let us analyze it a little. Three criticisms may be brought against it.

First, that it leads to hasty legislation without any adequate knowledge of the facts necessary to a wise conclusion and with a reckless disregard of all consequences except the political effect upon the party. It is perfectly true that hasty legislation is a characteristic of the modern tendency in politics. It is not, however, essentially connected with the caucus system; that is, the caucus system may facilitate such haste, but is not the cause for it. One reason for the desire to legislate immediately, without too careful consideration, is to be found in a natural reaction against the slow methods of legislation which have resulted from our constitutional system. The constitutional safeguards were originally adopted very largely to prevent a too hasty response to the immediate will of the people. Do not allow yourselves to be deluded by the phrases of some orators with the idea that a more direct response of the legislative body to the popular desire is a "restitution of the government

THE PARTY

to the people." It is not a question of restitution because the founders of the Republic carefully provided against hasty and impulsive action. If their judgment was in error we may adopt a new system, but remember that it will be a new system, not a return of something that has somehow been stolen from the people.

Unquestionably, under the old method it was frequently difficult for the people to secure the enactment of such measures as they desired and, furthermore, one reason for the revolt against the established system is the greater intensity of interest in the actual questions involved, to which I have referred above. The fact is, the people are impatient. They want immediate action. They frequently want action without adequate consideration of the complexity of the problem involved. As a result a party which is making an appeal to the people is inclined to declare itself immediately in some definite program with very little consideration of ultimate consequences so long as they meet the popular demand. Under party organization with approaching elections it is prob-

156 POLITICIAN, PARTY AND PEOPLE

ably too much to ask of human nature that the party leaders will not take such an attitude.

A year ago last November a large Democratic majority was elected to the House of Representatives. This was taken to mean on the part of most politicians that the people were dissatisfied with Republican rule and especially with Republican tariff policy. About a year ago Congress met in extraordinary session with the Democrats in control of the House and with a presidential campaign to be launched upon the country the following year. Under the circumstances it seemed more important to the Democratic leaders to declare some definite tariff policy at once rather than to work out carefully the details of a sane and well-rounded tariff measure on the basis of a thorough study of the facts. In fact, they introduced several measures of the most careless kind, hastily drawn, and put through the House under the caucus system. One reason for this, doubtless, was their certainty that such measures could not receive the approval of the President and therefore

THE PARTY 157

they need not worry much about details. I hardly need to tell you that legislation of this character seems to me wrong in principle and ultimately of grave danger to the country. I hope, however, that I am fair-minded enough to recognize that it was the result of the immediate political exigency and what to me seems the unreasoning impatience of the people themselves rather than to any fatal defect in the present legislative organization.

Theoretically, at least, under the system of the effective leadership of the few and the power of the caucus to secure results with certainty, legislation still might be carried on—and in the future I hope will prove to be carried on—in a more intelligent and thorough manner. The recent situation has been extremely strained and our political institutions should not be judged by these conditions alone. It would be possible even under the present system, especially with a more self-restrained voting population, for the leaders to take ample time for due consideration of all the many elements involved in a legislative program, to take time to

158 POLITICIAN, PARTY AND PEOPLE

secure the needed information as the basis of sane judgment, and to work out measures which, however much they might be opposed on principle by the other party, would stand the test of expert examination in the matters of detail. In the same way the caucus itself might, theoretically at least, become a place of genuine and effective debate as to the course of party policy.

The second objection is that of secrecy. The caucus is a private affair to which the public is not admitted, and it is true that, since the course of legislation is determined in the caucus, it does come in a measure to be legislating in secret. On the other hand, this secrecy cannot be preserved very inviolate and the action of the individual in the caucus can probably be learned by his constituents if they so desire. Some people have advocated an open caucus. It seems to me, however, that the closed caucus is more logical. One hardly would expect, for instance, the British cabinet to admit the public into their debates among themselves as to just what measures they would stand for or what particular form any measure is

THE PARTY 159

to take. If the opposition of this or that member were every time known in detail, the strength of the cabinet as a leading body would be diminished. Nor would we expect, for example, the Supreme Court to admit the outside public into its discussions of a case before decision had been rendered.

The third objection is that under the present system the minority has no influence whatsoever in the field of legislation. This, of course, is largely true. The majority practically agree beforehand that they will pay no attention to what the minority say. This certainly is a great change from the old theory of parliamentary government, that the leading minds of the country should get together and through frank discussion and interchange of opinion should arrive at a conclusion which represented the best opinion of the whole group. But under the present conditions, with the unwieldy character of the House, to which I have already referred, and the complexity of the problems involved, is not something of this kind practically essential to efficiency? After all, somehow Congress must legislate and those

160 POLITICIAN, PARTY AND PEOPLE

who are entrusted with the chief power must get measures through in some practicable way.

The result is, of course, in many ways unfortunate. It practically takes away the force of even the ablest speech in directing the course of legislation for the time being. The most powerful and eloquent debater on the minority side may make the effort of his life with about as little influence as King Canute had upon the waves of the ocean. Does this mean, however, that Congressional speeches are absolutely futile and that the power of effective debating is no longer an influence in determining the course of public affairs? Some people take this view. It is true, I think, that speeches are futile for the time being, and that bills are either passed or defeated without much regard for the speeches that are made on the floor. But they none the less have their ultimate purpose.

You see, the process is this. The bill is first framed by a group of leaders. It is secondly adopted by the caucus. It is thirdly defended on the floor of the House

THE PARTY 161

by the majority and attacked by the minority. The decision as to which argument is the better is not made at the time. This is already a foregone conclusion. But the decision as to which is the better policy is made at the polls at the next election and it is here that the arguments of the opposing sides will really count. Thus Congress becomes a great forum in which both sides of a question may be argued. The jury to make the ultimate decision consists of the people themselves and their verdict can only be rendered at a subsequent election. I am not sure that under present conditions this is not necessary for efficiency in legislation. It is all very different from our old theories of what a parliamentary body should be; it unquestionably makes us regret the passing of the old influence of debate; but it is practically essential under present conditions and one thing of the utmost importance which should be said in its favor is this; that it puts the responsibility for a given policy upon a given party and upon a given group of leaders in a most clear-cut manner.

I think this is a point which many people

have failed to realize—the great value which comes from a strict party responsibility. This is what I meant in suggesting earlier that we were evolving for ourselves a somewhat original type of responsible government. If measures of public importance were being passed always by a combination of voters on both sides, if neither one party nor the other as a whole stood absolutely in the lime light as responsible for this or that measure, the public would never know where to place responsibility and could probably be more easily hoodwinked than under the present system. This present system, carried out logically, means that the voters know which group of leaders carried through a certain group of measures. They can then judge the party as a whole according to its record. And they can take action accordingly.

I am perhaps forecasting the future somewhat too much. In the last twelve months we have seen this presumed efficiency of legislation rendered nugatory by a deadlock due to the fact that the President is of one party and the House of another. But this may always happen at any particular time, due to the difference in the length of tenure. I am looking on the subject, however, from the point of view of what will happen in the future when both branches of Congress and the executive position are controlled by the same party. It is, you see, all in line with what I said before regarding the position of the President. If the President is to become a party leader in the sense which I indicated, it is essential that when Congress is controlled by a majority of the other party he should stand by his own principles and by his own platform in all loyalty. Such a deadlock cannot long endure since the public will ultimately decide in favor of one party or the other. On the other hand, if this new power of the President is to continue, there will no longer be that instinctive deadlock between the President and Congress to which Mr. Hadley referred, provided that they are both of the same party. He and a group of sympathetic leaders in Congress will constitute an effective force for presenting to the people for their judgment a legislative policy which, however hastily and crudely framed in many ways, will be none the less definite and intelligible. And what is more, the country will hold them responsible for it.

What, then, becomes the position of the Congressman in the matter of loyalty to party under these new conditions?

I have already spent so much time attempting to explain the character of the party machinery for legislation that there is little time to go into detail into this question of the duty of the representative. You can see from what I have said that I am inclined to look with greater favor, even from a moral point of view, on party regularity than many public-spirited citizens with whose attitude you are familiar. This may be partly due to personal temperament, but it is more due, I think, to a conviction that in most cases it is today the only means of securing government efficiency. It doubtless has many evils, but I think in the long run that the worst of these are more than offset by the advantage which comes from party responsibility. That is one reason why I emphasized that point so strongly above.

THE PARTY 165

It is, of course, true that cases will arise where a representative cannot conscientiously stand by the action of his caucus. It may, for instance, be a matter of some fundamental political principle on which the individual member is absolutely unable, with a clear conscience, to yield his individual judgment. It is impossible here to take up detailed cases of such character. It might perhaps be such a case as the exemption of labor organizations from the operation of the anti-trust act, or from the use of injunctions against them. The individual member might consider this to be class legislation of such a character as to offend against the very foundation principles of American government and to the enactment of which he could under no circumstances be a party. Or it might be a proposition for some form of fiat money which, knowing that it would be destructive of commercial prosperity and even of national integrity, he could not possibly support. In such cases the honest legislator will state his case and refuse to be bound by the caucus rule. These are cases, however, where he would be ready to give up his political career altogether rather than to make any concession.

Such independence is possible to any representative today, where it is clearly a case of conscience, without losing him his standing in the party. He may even maintain a high position of leadership in the party after showing such independence in an individual case. Obviously, if he cannot conscientiously stand by the caucus in the case of a considerable number of leading party measures, he really does not belong in the party at all. It will not be so much a case of his being "read out of the party" as of his automatically leaving the party because unable to support its program.

But there are, however, a great many questions, possibly important ones, on which he may yield with a clear conscience on the ground that by insisting on voting according to his personal opinion he will ultimately do more harm than good. The harm would lie in disrupting that political machinery which is necessary for effective and responsible legislation.

Take, for example, an appropriation bill

THE PARTY

carrying appropriations for many different purposes, including, let us say, a dozen different bureaus and commissions of inquiry. Can the individual vote on each one solely according to his opinion of its desirability? Can even a member of the Appropriations Committee itself make a lone fight, involving waste of time and dangerous friction, against what he knows to be the overwhelming sentiment of his party? Especially can he do this when he knows that the members of the opposing side are not voting according to personal conviction, but as a unit to embarrass or disrupt his side as far as possible? Various friends have frequently expressed to me surprise at the attitude of some particular Congressman on some particular question of minor inportance. When I have explained that individually the member was (say) in favor of the proposition, but voted against it because of party necessity, the disgusted answer has frequently been, "Is that the kind of men we send to Congress?" I think such an attitude is entirely unfair toward many of the most conscientious and far-sighted of our representatives. Unless it is a matter that seems to him to vitally affect our principles of government, or the very foundations of our prosperity, he is justified in yielding his individual opinion. It is not merely a service to party. It may be a service to the country in the double sense that it increases the efficiency of government and also makes more clear-cut the party's responsibility to the public.

It is true that the individual Congressman may take the attitude that he will vote always for every measure according to whether he thinks it desirable or undesirable in itself, regardless of any party. In such case, however, he separates himself from all parties and his influence is largely destroyed. He no longer can make himself felt in those meetings where policies are really determined. He can only cast his vote and make his lone protest.

This, you see, is the second great problem of the Congressman, the first having been the question as to how far he shall be loyal to the interests of his own section as against what he considers the general good. Not

THE PARTY

infrequently the question of these two duties comes into a serious clash. Obviously, the man who breaks from party loyalty for the sake of the interests of his own constituents is doing so on a much lower ground than he who does it because of his loyalty to some great principle of government. In most cases such action will be due simply to his fear of his constituents; that is, his fear that he will fail of re-election. Of course, it may be that he honestly believes the party policy is destructive of the welfare of his community and that to block it he would be willing not only to break with the party, but to retire from public life altogether. This, I think, is the real standard of conscientious action. To vote for a measure merely in order to stay in can hardly receive the commendation of any right-thinking man. To vote for a measure when, in order to secure its passage, a man is willing to get out, is a practice which no one could condemn.

Commonly, in the conflict between his two loyalties, the representative will make his fight in the caucus in behalf of his constituents. If he is unable to change the party

170 POLITICIAN, PARTY AND PEOPLE

policy, he will usually sacrifice what he considers the interests of his constituents to the maintenance of the coherence of his party. In doing so I think he is following a higher duty. For, after all, the party is a national party representing the whole country and responsible to it for its actions. There have been conspicuous cases where men have remained loyal to the party program, while frankly announcing that by so doing they were destroying every chance of their own continuance in public life. In extreme cases the representative may sacrifice his loyalty to party to his loyalty to section. In doing so he is almost certain to lose such influence in party councils as he may have had and, what is more, it is doubtful if he will win the real respect of his constituents in the end.

I told you at the outset that I should probably make your problems difficult rather than easy; that I should propound problems for you rather than solve them for you. You see that I have done so. I fear, too, that you may think that I have spoken from a somewhat low moral plane in suggesting that a conscientious man must really

THE PARTY

confront problems of this kind. It would have been easy to simply tell you that you should always vote for what is right, but I warned you in the beginning that the most difficult problem is to find out what is right. The conditions are so intricate that there is no single rule which can be laid down to guide a man's conduct. True, one may say that he should always act for what he thinks is the general good in the long run. To that I would subscribe as heartily as anyone else. It is the very fact, however, that the most moral man may conscientiously believe that yielding his own opinion for the time being may work for the greater good in the end which makes the decision difficult.

It would have been easy to hold up to you some ideal of a body of patriotic, independent legislators having no personal interest, no sectional interest, and no party interest, but however inspiring such an address might have been made, it would have done little to illuminate the problems which you will have to face as you go into the world as it is. An old German friend of mine used to remark serenely, "Man muss mit Tatsachen rechnen und das Unvermeidliche mit Würde tragen"—one must reckon with the facts and bear the inevitable with dignity. It is, I think, a higher duty to face inevitable facts as they are and then to strive conscientiously to work toward the best results within the limitations which these facts impose.

I also told you at the beginning of these lectures that your first duty was knowledge and I come back to that as your final duty. You must take these matters seriously; you must study them; you must ponder over them. You must study not only what policy would be best if you were a despot and could decide the matter alone. You must study also the facts of political life, the facts of human nature, the problems of what can actually be accomplished, as well as what you would like to see accomplished.

Again, as I said at first, I have spoken more strongly on one side than I might otherwise have done, because of the character of my audience. It is because I have felt that it is young men of your type who are least tolerant in such matters and most likely to forget the necessity of efficiency in action, while glorifying the importance of your own personal will and opinion. There is a poem by Edward Rowland Sill, entitled "Dare You?" which puts this problem well. I think it is not misusing it to apply it even in the political field. "Doubting Thomas" says to "loving John":

> Tell me now, John, dare you be
> One of the minority
> To be lonely in your thought
> Never visited or sought?
> If you dare, come now with me,
> Fearless, confident, and free.

To this John replies:

> Thomas, do you dare to be
> Of the great majority?

The poet would suggest that sometimes it takes a higher courage to sink one's own individuality of thought and action in the cause of some higher "unity." In all walks of life the problem of when and to what extent this should be done will confront you, puzzling and recurring; nowhere more than in the field of politics.

INDEX

Action, more effective than discussion for good government 53, 58-59
Administration, local, compared with national 64-65
"Agency theory" 113-115, 119, 130
Aldrich, Senator 138
Allison, Senator 138
Ambition in politics 88-89
American politics (Nineteenth Century), compared with Eighteenth Century English (Hadley) 139-141
Appointments, government 124-126
Appropriations, securing 126, 167
Arbitration in industrial disputes 111
Averaging interests of localities 107-108
"Big Five" in United States Government 138
Bills, legislative, process of formation and enactment, 160-161
Blaine, James G., 91; quoted, 92.
"Boss," unreasonable prejudice against, 57-59; importance, 136.
British cabinet 158
Burke, Edmund, quoted..................103-106, 116-118
Business ethics, lectures on 1
Cabinet government 142
Candidate, should be chosen according to leader he will obey 67-68
Catholics 40
Caucus, party, — policies framed and amended in, 152-153; — debated upon in, 158; — growth in power, 153-154; — analyzed and criticised, 154-155; — open or closed, 158; — secret legislation in, 158; — constituents' interests defended in, 170.
Claims, of *country,* as against *party* and *district* ...101-102
Class legislation 165-166
Class- or group-interest party 39-40

INDEX

Clearing house of information on matters of public policy needed 28-29
Commercial matters, interference of legislator in......109
Commission system of government 70
Common welfare, how best achieved 6-7
Compromise in legislation often necessary 76-77
Confidence in representatives needed 89-92
Congress, a forum for argument 161
Congressional representatives, relation to the President 80-81, 145
Congressional votes, their importance 71
Congressmen, — should they work solely for interests of constituents? 102-103, 120-121; — bravery in right action, 121; — concerned with federal affairs in their districts, 121-127; — tariff questions before, 132-133.
Congressmen. See also *Representatives, Congressional.*
Conscience and judgment in choice of party..........42-43
Constituents, — claims of, as against those of country, 104-108, 114, 120-129; — great variety of interests, 116-120; — duty of representative in following opinions of, 104-106; — interests defended in caucus, 170; loyalty of representatives to, — when wrong, 169.
Constitutional system in United States Government causes deadlocks 135-136, 140-141, 162-163
Continuance of service more important than individual originality in statesmen 96-99
Contracts, government, awarding of 127
Cynicism among representatives engendered by suspicion 86
Deadlocks, occurrence in United States Government, 135-136, 140-141, 162-163
Debate in Congress, no longer effective? 160-161
Democratic Congress elected 156
Democratic Party, — in South, 39; — and Republican Party, 40.

INDEX

Democratic tariff policy 156
Departments, administrative, extravagance in......129-130
Direct legislation 111-113
Direct primaries 111-112
Disraeli, Earl of, quoted 151
Earnestness in political life, new spirit in United
  States 147-149
Economy, difficult to maintain by Congress 127-129
Editors, — large following, 37; — less trustworthy than
  supposed, 14.
Elections, local, — should be separate from national in
  time of holding, 45, 64-65, 69; — possible bearing
  on national elections, 68-69.
England, her King's position 146
English politics (Eighteenth Century), compared with
  Nineteenth Century American (by Hadley) 139-141
Ethical viewpoint of undergraduates 2-5
Ethics of politics, Chapter I.
Extravagance in Government Departments 129-130
Fact, determination of, very difficult to voter 29-31
Facts and the Voter, Chapter I.
Fairness to representatives needed 94
Familiarity breeds contempt in politics 91-94
Foundation for present lectures 1
Free press, greatest safeguard of democratic govern-
  ment 13
Germany, class-interest parties in 39
Government and legislation, matters of reason and
  judgment (Burke) 104-105
Hadley, President A. T., quoted ..101-102, 135-136, 139-140
Hale, Senator 138
Headlines, important influence 19-21
Historical study needed for successful legislation 9
Honesty in politics 89-91, 94
House of Representatives, growing size, consequences,
  142-144
Immediate results, not always possible in politics 95

INDEX

Immoral voting 79-80
Independence, — in voting, 43-46; — in Congress, 168-169; — in representatives' judgment, 105-107, 114-115; — of party, 55-56.
Independence of party, overemphasized by college men, 55-56
Independent action, not always advisable for legislators, 83
Individual candidates *vs.* strict party loyalty, 61-64, *et seq.;* 78-79.
Individualism 107-11, 173
Industrial disputes, arbitration in 111
Information, accurate, difficult to obtain, Chapter I.
Information on public policy, clearing house needed..28-29
Initiative and referendum 35-36, 112
Interviews, misrepresented in press 16
Judgment, hasty, really immoral 11-12
King, his position in England 146
Kipling, Rudyard, quoted 94
Knowledge, voter's first duty, Chapter I; 172
Laboring class, studious of political questions 7-8
Laws, good, spoiled by poor framing and enforcement...63
Leaders, intellectual 37-38
Leaders, — political, intelligent choice of, the voter's first duty, 31-32, 34-35, 37-38; — choice of, means choice of party, 38-39; — their importance, 66-67, 78; — relative power as compared with relative insignificance of rank and file, 74; compromises often misunderstood, 77; displacement of old leaders in United States, 147.
Leadership, political, psychology of 76-77
Legislation, — necessity of historical study for successful, 9; — enforcement as important as enactment, 62-63; — misdirected, 9.
Legislation, class 165-166
Legislation, direct, — when successful, 36; — dangers in, 36, 111-113.
Legislation, "private," in Congress 119

INDEX 179

Legislation, hasty 154-156
Legislation, secret, in caucus 158
Legislative Reference Library in Wisconsin 29
Legislature, of what it must consist 66
Lincoln as a leader 77
Local administration compared with national 64-65
Loyalty, party, 43-53; — misunderstood, 54-55, 167-168; increased necessity for, 101-102, 134; Chapter V (p. 135 *et seq.*); — supplanting loyalty to district, 150-151; — among Congressmen, 164-166, 169-170.
Loyalty of representatives to constituency, when wrong, 169
Loyalty of constituency to elected representatives needed 99
"Machines," local, circumstances tending to continued power 68-69
Maine, prohibition in 42
Majority rule in United States legislation 159-161
Majority vote, inevitably for best interests of whole? 109-110
Malthusian theory 5
Mathematical principles applied to politics 110-112
Measures and men, go together 63-64
"Measures not men," 61; — best policy in Congressional elections, 71-75.
Men, rather than measures, in *local* politics 64-65
Middle West, influence in politics 142
Minority, no voice in legislation 159
Money in politics 87
Motives in politics 87-89
Municipal and other local elections 64-65
Names, party, should be different in local and national affairs 69-70
News, voter should compare and discriminate in 26-28
News and "facts," difference 16-17
News, incorrect, seldom retracted, — bad influence ..22-25

INDEX

Newspaper correspondents, — trustworthy, 14; — often ignorant of subject, 15-16.

"One man power" 144

Page, Edward D. 1

Parliament, a, defined by Edmund Burke..... 102, 105-106

Parties, — political, in United States, usually two great parties, 40-41; — why smaller ones attract few adherents, 41; — three great groups, 38-40; — without *any* policy for public good, 54.

Party, the, and the Voter, Chapter II.

Party, difficulty of right choice, 42-43, 55-56; relation of representative to, 135, *et seq.*

Party divisions the same for local, state, and national affairs at present, 68; unfortunate effect of this, 68-69.

Party government and representative government, 33-34; — in United States, 147-148.

Party politics, national, differentiated from state and local 44-45

Party loyalty.... 43-55, 101-102, 134-135, 150-151, 164-170

Party measures, not paramount in local administration, 64

Party names, should differ in local and national affairs, 69-70

Party policies, by whom determined 75-78

Party responsibility 162-164

Patronage, a difficult problem for Congressmen 123-126

Payne-Aldrich Tariff Act of 1909 29-31

Peel, Sir Robert 151

Pension bills, private 119-120

Periodical literature, influence powerful, 11; comment prejudiced, 25-26.

Periods in party government in United States 147-148

Personal friendship in voting 79

Personalities in United States Government, change in.. 138

Personality in politics 38, 61-64

Platt, Senator 138

Pliancy necessary to good political leadership 77-78

INDEX 181

Policies (Presidential), a new element in United States Government 143-144
Policy, party, determined by leaders or rank and file 75-78
Political economy, so-called *law* not moral force 4-5
Political leadership, psychology of 76-77
Politicians, generally suspected 85-87
Politicians and *statesmen* 85
Politics, ethical side 1-4
Politics, — right and wrong in, hard to determine, 9-11; — an honorable career, 89-90.
"Pork-barrel," the 126
President, — free from sectional control, 147; — increasing power of, 139-145, 163; — relations to Congress, 80-81, 145; — unique position politically, 145-146.
Press, — injustice done public men in, 11-19; — our opinions gained from, 13; — power of, 11-13.
Primaries, direct 111-112
Principle, change of, in the United States Government, 138-142
"Private" legislation in Congress 119
Problems propounded, not solved, in these lectures.....171
Prohibition, in Maine 42
Prohibition Party 41
Promises, party 155
Punishment of individual occasional necessity 70, 83
Recall 112
Reed, Thomas B., — definition of a statesman, 85; — quoted, 110; — able to be independent of constituents' opinions, 115.
Reporters (newspaper), ethics high 14-16
Representative, Congressional, — choice of, 61-62, 70-72; — factors in choice, 73-75; — relations to President, 80-81, 145.
Representatives, Congressional. See also *Congressmen.*
Representative, — and his constituency, Chapter IV (p. 100 *et seq.*); — duties to constituents, defined

182 INDEX

by Burke, 103-106, 117-118; — independence of judgment, 105-107, 114-115, 153; — controlled by party, 114; — variety of claims of constituents, 115-116; — independence of judgment, when advisable, 166.

Representative government and party government....33-34
Representative government, is it satisfactory? 35-36
Republican and Democratic Parties, principal difference 40
Republican Party in Maine 42
Republican tariff policy 156
Responsibility for legislative policies fixed 161-164
"Restitution of the government to the people"......155-156
Results, immediate, not always possible in politics 95
Right and wrong in politics, a complex question, Chapter I.
Roosevelt, Theodore, "ruler" of United States 137; 143
"Rotation in office" a bad thing 97-99
"Rulers" in America, who are they? 137-139
Secrecy in caucuses objectionable 158
Sectional-interest party group 39
Seniority in Congressional Committee selection 96
Separation of powers in government, extreme 141
Service, long, more important than originality in statesman 96-99
Sill, E. R., quoted 173
Smith, Adam, quoted 109
Socialist Party 42-43
South, Democratic Party in 39
"Splitting the ballot," is it advisable? ....67 *et seq.*, 81-84
"Spoils System" 124-125, 148
Spooner, Senator 138
State elections, duty of voters in, compared with municipal and national elections 65-66
State politics, vary with the States 66
System, old, breakdown of in United States Government 141-144

INDEX

Taft, Wm. H. ....,.................... 143
Tariff, varied bearing on different districts, 131-132;
— compromises in, 132-133.
Temperament in choice of party 55
Tenure of office ....140 *et seq.*, 163
Tenure of Office. See also *Continuance of Service.*
Tools, political, choice of 60
Tweed 140
Vote, result nullified by too independent choice, 67; 81-82
Voter, the, — and the Party, Chapter II; — and the
  Representative, Chapter III 61
Voter, — ethical duty of, 8-10; — limitation of his
  action, 43-44; —independence of party, 44; — conflicting duties of in municipal, state, and national
  elections, 68; — duty to representative after election, 84 *et seq.*
Voter, educated, ignorance and indifference to public
  questions 7
Voting, immoral 79-80
Walpole, Horace 140
Watchfulness of public servants needed 89-90
Wisconsin, Legislative Reference Library in 29
Young men in English politics 95-96
Young statesmen, their opportunities in United States
  politics 95-96